EVERY STUDENT CAN LEARN MATHEMATICS

# MATHEMATICS STRATEGIES FOR Tier 1 & Tier 2 Interventions

in a PLC at Work®

Sarah Schuhl

Mona Toncheff

Jennifer Deinhart

Brian Buckhalter

*foreword by Mike Mattos*

Solution Tree | Press

A Joint Publication With

Copyright © 2025 by Solution Tree Press

Materials appearing here are copyrighted. With one exception, all rights are reserved. Readers may reproduce only those pages marked "Reproducible." Otherwise, no part of this book may be reproduced or transmitted in any form or by any means (electronic, photocopying, recording, or otherwise) without prior written permission of the publisher.

555 North Morton Street
Bloomington, IN 47404
800.733.6786 (toll free) / 812.336.7700
FAX: 812.336.7790

email: info@SolutionTree.com
SolutionTree.com

Visit **go.SolutionTree.com/MathematicsatWork** to download the free reproducibles in this book.

Printed in the United States of America

Library of Congress Cataloging-in-Publication Data

Names: Schuhl, Sarah, author. | Toncheff, Mona, author. | Deinhart, Jennifer, author. | Buckhalter, Brian, author.
Title: Mathematics strategies for tier 1 and tier 2 interventions in a plc at work / Sarah Schuhl, Mona Toncheff, Jennifer Deinhart, Brian Buckhalter.
Other titles: Mathematics strategies for tier one and tier two interventions in a PLC at work
Description: Bloomington, IN : Solution Tree Press, [2025] | Includes bibliographical references and index.
Identifiers: LCCN 2024033006 (print) | LCCN 2024033007 (ebook) | ISBN 9781962188357 (paperback) | ISBN 9781962188364 (ebook)
Subjects: LCSH: Mathematics--Study and teaching. | Professional learning communities.
Classification: LCC QA11.2 .S38 2025  (print) | LCC QA11.2  (ebook) | DDC 510.71--dc23/eng20240927
LC record available at https://lccn.loc.gov/2024033006
LC ebook record available at https://lccn.loc.gov/2024033007

---

**Solution Tree**
Jeffrey C. Jones, CEO
Edmund M. Ackerman, President

**Solution Tree Press**
*President and Publisher:* Douglas M. Rife
*Associate Publishers:* Todd Brakke and Kendra Slayton
*Editorial Director:* Laurel Hecker
*Art Director:* Rian Anderson
*Copy Chief:* Jessi Finn
*Senior Production Editor:* Christine Hood
*Copy Editor:* Jessica Starr
*Proofreader:* Elijah Oates
*Cover and Text Designer:* Laura Cox
*Acquisitions Editors:* Carol Collins and Hilary Goff
*Content Development Specialist:* Amy Rubenstein
*Associate Editors:* Sarah Ludwig and Elijah Oates
*Editorial Assistant:* Madison Chartier

# Acknowledgments

We have the honor of working with collaborative mathematics teams across the United States using the Mathematics in a PLC at Work® process. Our work from the *Every Student Can Learn Mathematics* series supports collaborative mathematics teams with the following.

1. Planning instructional units by making sense of standards, mathematical tasks, and pacing
2. Creating common assessments and using them for teacher and student learning
3. Designing high-quality lessons with intentional instructional routines
4. Exploring grading as feedback

We would like to express our gratitude to the educators who challenged us to consider how we should rigorously teach mathematics—at grade level and beyond—to every student through targeted, connected, and meaningful instruction in Tier 1 and Tier 2 learning experiences. This book was developed to share mathematics strategies that you and your teams can use while addressing the four critical questions of a PLC at Work (DuFour et al., 2024).

We are grateful to Jeff Jones, Douglas Rife, Todd Brakke, Rian Anderson, and Christine Hood, along with the editorial team, for understanding our vision for Mathematics in a PLC at Work. Thank you for supporting this additional resource, which is intended to deepen the way collaborative teams make sense of, teach, assess, and design interventions for mathematics. Thank you also to our reviewers and colleagues for your feedback and insights.

We owe a deep debt of gratitude to our families, who support us as we learn from one another and travel to work with mathematics teams across the United States. Thank you for gifting us the time to write this book.

Finally, we extend our thanks to you, our readers, for the work you are doing in schools to ensure every student learns mathematics. You and your collaborative teams labor tirelessly to provide all students with meaningful and equitable mathematics learning experiences. You are committed to accelerating the learning of students not yet at grade level, while simultaneously challenging those who are learning at grade level to deepen their understanding of mathematics. Your work embodies the series title that this book is a part of: *Every Student Can Learn Mathematics.*

Solution Tree Press would like to thank the following reviewers:

Kristy Scott Giplin
Assistant Principal
Rollins Place Elementary School
Zachary, Louisiana

Gwen Savario
First-Grade Teacher
Rollins Place Elementary School
Zachary, Louisiana

Candies Winfun-Cook
Clinical Assistant Professor
School of Education, Teacher Education Department
The University of Mississippi
Oxford, Mississippi

Julie James
Assistant Director of Professional Learning
Center for Mathematics and Science Education
The University of Mississippi
Oxford, Mississippi

Jason Cianfrance
Mathematics Teacher
Legacy High School
Broomfield, Colorado

Tracey A. Hulen
Author, Education Consultant, Mathematics and
 SEL Specialist
T. H. Educational Solutions, LLC
Fairfax, Virginia

Ashley Mistretta
Instructional Coach
Rose Hill Elementary School
Fairfax County Public Schools
Alexandria, Virginia

Janet Nuzzie
District Intervention Specialist, K–12 Mathematics
Pasadena Independent School District
Pasadena, Texas

Gwen Zimmermann
Educational Consultant
Stanardsville, Virginia

---

Visit **go.SolutionTree.com/MathematicsatWork** to download the free reproducibles in this book.

# Table of Contents

*Reproducibles are in italics.*

**About the Authors** . . . . . . . . . . . . . . . . . . . . . . . . . . . . . . . . . . . . . . . . . . . . . . . . . . . . . . . . . . . . . . . xi

**Foreword** . . . . . . . . . . . . . . . . . . . . . . . . . . . . . . . . . . . . . . . . . . . . . . . . . . . . . . . . . . . . . . . . . . . . . . . . . xv

**Introduction** . . . . . . . . . . . . . . . . . . . . . . . . . . . . . . . . . . . . . . . . . . . . . . . . . . . . . . . . . . . . . . . . . . . . . . 1

    The Current Reality . . . . . . . . . . . . . . . . . . . . . . . . . . . . . . . . . . . . . . . . . . . . . . . . . . . . . . . . . 1

    Mathematics in a PLC at Work . . . . . . . . . . . . . . . . . . . . . . . . . . . . . . . . . . . . . . . . . . . . . . 2

    What's in This Book . . . . . . . . . . . . . . . . . . . . . . . . . . . . . . . . . . . . . . . . . . . . . . . . . . . . . . . . 4

        Part 1: A Culture of Learning . . . . . . . . . . . . . . . . . . . . . . . . . . . . . . . . . . . . . . . . . . 4

        Part 2: Mathematics Foundations . . . . . . . . . . . . . . . . . . . . . . . . . . . . . . . . . . . . . 4

        Part 3: Student Engagement . . . . . . . . . . . . . . . . . . . . . . . . . . . . . . . . . . . . . . . . . 5

    Conclusion . . . . . . . . . . . . . . . . . . . . . . . . . . . . . . . . . . . . . . . . . . . . . . . . . . . . . . . . . . . . . . . . 5

## PART 1: A CULTURE OF LEARNING . . . . . . . . . . . . . . . . . . . . . . . . . . . . . . . . . . . . . . . . . . 7

### 1   Work as a Collaborative Mathematics Team Focused on Student Learning . . . . . . . 9

The *Why* and the *What* of Working as a Collaborative Mathematics Team Focused on Student Learning . . . . . . . . . . . . . . . . . . . . . . . . . . . . . . . . . . . . . . . . . . . . . . . . . . . . . . . . . . . . . . . . . . . . 9

    Tier 1 Preventions . . . . . . . . . . . . . . . . . . . . . . . . . . . . . . . . . . . . . . . . . . . . . . . . . . . . . . . 11

    Tier 2 Interventions and Extensions . . . . . . . . . . . . . . . . . . . . . . . . . . . . . . . . . . . . . . 12

    Tier 3 Intensive Reinforcements . . . . . . . . . . . . . . . . . . . . . . . . . . . . . . . . . . . . . . . . . 13

The *How* of Working as a Collaborative Mathematics Team Focused on Student Learning . . . . . . . . . . . . . . . . . . . . . . . . . . . . . . . . . . . . . . . . . . . . . . . . . . . . . . . . . . . . . . . . . . . . 13

　　　　Time in the School Day for Tier 2 Interventions and Extensions . . . . . . . . . . . . . . . . . 13
　　　　Tier 1 and Tier 2 Instruction on Essential Learning Standards . . . . . . . . . . . . . . . . . 14
　　　　Rigorous Mathematics. . . . . . . . . . . . . . . . . . . . . . . . . . . . . . . . . . . . . . . . . . . . . . . 15
　　　　High-Quality Tier 1 and Tier 2 Instruction. . . . . . . . . . . . . . . . . . . . . . . . . . . . . . . . 17
　　Conclusion . . . . . . . . . . . . . . . . . . . . . . . . . . . . . . . . . . . . . . . . . . . . . . . . . . . . . . . . . . . 22
　　Questions to Consider for Next Steps . . . . . . . . . . . . . . . . . . . . . . . . . . . . . . . . . . . . . 22

## 2  Build a Community of Learners　　　　　　　　　　　　　　　　　　　　27

　　The *Why* and the *What* of Building a Community of Learners . . . . . . . . . . . . . . . . . . 27
　　The *How* of Building a Community of Learners . . . . . . . . . . . . . . . . . . . . . . . . . . . . . 30
　　　　Norms for Student Interactions . . . . . . . . . . . . . . . . . . . . . . . . . . . . . . . . . . . . . . . 30
　　　　Learning Community Environment . . . . . . . . . . . . . . . . . . . . . . . . . . . . . . . . . . . 33
　　　　Mistakes as Learning Tools . . . . . . . . . . . . . . . . . . . . . . . . . . . . . . . . . . . . . . . . . 33
　　Conclusion . . . . . . . . . . . . . . . . . . . . . . . . . . . . . . . . . . . . . . . . . . . . . . . . . . . . . . . . . . . 38
　　Questions to Consider for Next Steps . . . . . . . . . . . . . . . . . . . . . . . . . . . . . . . . . . . . . 39

# PART 2: MATHEMATICS FOUNDATIONS . . . . . . . . . . . . . . . . . . . . . . . . . . 41

## 3  Teach Grade- or Course-Level Content　　　　　　　　　　　　　　　　43

　　The *Why* and the *What* of Teaching Grade- and Course-Level Content . . . . . . . . . 43
　　The *How* of Teaching Grade- or Course-Level Content . . . . . . . . . . . . . . . . . . . . . . 45
　　　　Essential Standards. . . . . . . . . . . . . . . . . . . . . . . . . . . . . . . . . . . . . . . . . . . . . . . . 45
　　　　High-Level-Cognitive-Demand Tasks . . . . . . . . . . . . . . . . . . . . . . . . . . . . . . . . . 46
　　　　Leveled Tasks. . . . . . . . . . . . . . . . . . . . . . . . . . . . . . . . . . . . . . . . . . . . . . . . . . . . 49
　　Conclusion . . . . . . . . . . . . . . . . . . . . . . . . . . . . . . . . . . . . . . . . . . . . . . . . . . . . . . . . . . . 54
　　Questions to Consider for Next Steps . . . . . . . . . . . . . . . . . . . . . . . . . . . . . . . . . . . . . 54

## 4  Connect to Prior Knowledge　　　　　　　　　　　　　　　　　　　　　57

　　The *Why* and the *What* of Connecting to Prior Knowledge . . . . . . . . . . . . . . . . . . . 57
　　The *How* of Connecting to Prior Knowledge . . . . . . . . . . . . . . . . . . . . . . . . . . . . . . . 59
　　　　Preassessments. . . . . . . . . . . . . . . . . . . . . . . . . . . . . . . . . . . . . . . . . . . . . . . . . . . 59
　　　　Just-in-Time Supports . . . . . . . . . . . . . . . . . . . . . . . . . . . . . . . . . . . . . . . . . . . . . 60
　　　　Learning Progressions . . . . . . . . . . . . . . . . . . . . . . . . . . . . . . . . . . . . . . . . . . . . . 61
　　　　Prior Knowledge Routines. . . . . . . . . . . . . . . . . . . . . . . . . . . . . . . . . . . . . . . . . . 61
　　　　Centers and Games. . . . . . . . . . . . . . . . . . . . . . . . . . . . . . . . . . . . . . . . . . . . . . . . 66
　　Conclusion . . . . . . . . . . . . . . . . . . . . . . . . . . . . . . . . . . . . . . . . . . . . . . . . . . . . . . . . . . . 67
　　Questions to Consider for Next Steps . . . . . . . . . . . . . . . . . . . . . . . . . . . . . . . . . . . . . 67

## 5 Develop Number Sense ... 69

The *Why* and the *What* of Developing Number Sense ... 69

The *How* of Developing Number Sense ... 71

- Early Numeracy ... 71
- Subitizing Routines ... 72
- Forward and Backward Counting Strategies ... 74
- Number Lines ... 76
- Number Talks ... 78
- Estimation and Reasonable Answers ... 78

Conclusion ... 80

Questions to Consider for Next Steps ... 80

## 6 Focus on Problem Solving ... 83

The *Why* and the *What* of Focusing on Problem Solving ... 83

The *How* of Focusing on Problem Solving ... 85

- Visuals and Tools ... 87
- Problems Without Numbers ... 91
- Three Reads Protocol ... 91

Conclusion ... 92

Questions to Consider for Next Steps ... 92

## 7 Develop Procedural Fluency ... 95

The *Why* and the *What* of Developing Procedural Fluency ... 95

- CRA Method in Tier 1 ... 97
- CRA Method in Tier 2 ... 97

The *How* of Developing Procedural Fluency ... 99

- Flexible Strategies Versus Memorization ... 99
- Known Ten Strategy ... 100
- Vertical Progression of Strategies ... 101
- Existing Pictorial Models ... 104
- Strategy Connections and Comparisons ... 104
- Just Right Numbers ... 105
- Over or Under Estimation ... 106
- My Favorite Mistake ... 106
- Card Sort ... 107

Conclusion ... 109

Questions to Consider for Next Steps ... 109

## PART 3: STUDENT ENGAGEMENT ... 113

### 8 Communicate Using Mathematical Language ... 115
The *Why* and the *What* of Communicating Using Mathematical Language ... 115
The *How* of Communicating Using Mathematical Language ... 118
- Word Walls ... 119
- Which One Doesn't Belong? ... 120
- Would You Rather? ... 120
- Share-Trade Protocol and Stronger and Clearer Each Time ... 122
- Mathematics Journaling ... 123
- Consensus Boards ... 124
- Mathematics Graphic Organizers ... 124

Conclusion ... 125
Questions to Consider for Next Steps ... 125

### 9 Grow Learning Through Student Discourse ... 129
The *Why* and the *What* of Growing Learning Through Student Discourse ... 129
The *How* of Growing Learning Through Student Discourse ... 132
- Purposeful Questioning ... 132
- Sentence Frames ... 133
- Flexible Grouping ... 135
- Jigsaw ... 136

Conclusion ... 137
Questions to Consider for Next Steps ... 137

### 10 Use Meaningful Feedback for Learning ... 139
The *Why* and the *What* of Using Meaningful Feedback for Learning ... 139
The *How* of Using Meaningful Feedback for Learning ... 142
- Show the Mathematics ... 142
- Three Es ... 145
- Provide the Answers ... 146
- Learning Teams ... 148
- Task Sorting ... 148
- Highlighters ... 148
- Student Questions ... 149
- Error Analysis ... 149
- Learning Goals and Reflection ... 150

Conclusion ... 150
Questions to Consider for Next Steps ... 150

## 11 Empower Learners Through Student Investment ... 153

The *Why* and the *What* of Empowering Learners Through Student Investment ... 153

The *How* of Empowering Learners Through Student Investment ... 156

- Goal Cards ... 156
- Goal Setting ... 158
- Rubrics ... 158
- Success Criteria ... 161
- Proficiency Scales ... 163
- Student Trackers ... 165
- Student Surveys ... 165

Conclusion ... 168

Questions to Consider for Next Steps ... 169

## Epilogue ... 171

## Appendix A: Data Analysis Protocols ... 173

*Student Work Protocol* ... 175

*Essential Learning Standard Analysis Protocol: Grade 4 Sample* ... 176

## Appendix B: Cognitive-Demand-Level Task Analysis Guide ... 179

*Cognitive-Demand Levels of Mathematical Tasks* ... 180

## Appendix C: Team Actions to Avoid and Consider for Tier 1 and Tier 2 ... 181

*Team Actions to Avoid and Consider for Tier 1 and Tier 2* ... 182

## References and Resources ... 183

## Index ... 191

# About the Authors

**Sarah Schuhl** is an educational coach and international consultant specializing in mathematics, professional learning communities (PLCs), common formative and summative assessments, priority school improvement, and response to intervention (RTI). She has worked in schools as a secondary mathematics teacher, high school instructional coach, and K–12 mathematics specialist.

Schuhl was instrumental in the creation of a PLC in the Centennial School District in Oregon, helping teachers make large gains in student achievement. She earned the Centennial School District Triple C Staff Recognition Award in 2012.

Schuhl fosters learning through large-group professional development and small-group coaching in districts and schools. Her work focuses on strengthening the teaching and learning of mathematics, having teachers learn from one another when working effectively as a collaborative team in a PLC, and striving to ensure every student learns through assessment practices and intervention. Her practical approach includes working with teachers and administrators to implement assessments for learning, analyze data, collectively respond to student learning, and map standards.

For Mathematics at Work, Schuhl coauthored *Engage in the Mathematical Practices: Strategies to Build Numeracy and Literacy With K–5 Learners*, as well as the *Every Student Can Learn Mathematics* series and the *Mathematics at Work Plan Book* with Timothy D. Kanold. She was also editor of the *Mathematics Unit Planning in a PLC at Work* series and contributed to the National Council of Supervisors of Mathematics (NCSM) publication *NCSM Essential Actions: Framework for Leadership in Mathematics Education*.

Schuhl coauthored *Acceleration for All: A How-To Guide for Overcoming Achievement Gaps* and *School Improvement for All: A How-To Guide for Doing the Right Work* about priority schools. She contributed to *Charting the Course for Leaders: Lessons From Priority Schools in a PLC at Work* and *Charting the Course for Collaborative Teams: Lessons From Priority Schools in a PLC at Work*.

Previously, Schuhl served as a member and chair of the National Council of Teachers of Mathematics (NCTM) editorial panel for the journal *Mathematics Teacher* and as secretary of NCSM. Her work with the Oregon Department of Education includes designing mathematics assessment items, test specifications and blueprints, and rubrics for achievement-level descriptors. She has also been a contributing writer for a middle school mathematics series and an elementary mathematics intervention program.

Schuhl earned a bachelor of science in mathematics from Eastern Oregon University and a master of science in mathematics education from Portland State University.

To learn more about Sarah Schuhl's work, follow her @SSchuhl on X.

**Mona Toncheff** is an educational consultant and author with more than thirty years of experience in public education. She worked as both a mathematics teacher and a mathematics specialist for the Phoenix Union High School District in Arizona. In the latter role, she coached and provided professional development to high school teachers and administrators related to quality mathematics teaching and learning and effective collaborative teams. She currently serves as a supervisor teacher for the University of Arizona Teach Arizona Program.

Toncheff has supervised the culture change from teacher isolation to professional learning communities; created articulated standards and relevant district common assessments; and provided ongoing professional development on best practices, equity and access, technology, response to intervention, high-quality grading practices, and assessment for learning.

Toncheff has coauthored several Solution Tree and NCSM publications. Her Solution Tree books include *Common Core Mathematics in a PLC at Work, High School*; *Beyond the Common Core: A Handbook for Mathematics in a PLC at Work, High School*; and *Activating the Vision: The Four Keys of Mathematics Leadership*. She is also one of the lead coauthors of the *Every Student Can Learn Mathematics* series. Toncheff was the lead editor and contributing author for the NCSM books *Framework for Leadership in Mathematics Education*, *Instructional Leadership in Mathematics Education*, and *Culturally Relevant Leadership in Mathematics Education*.

As a writer and consultant, Toncheff works with educators and leaders across the United States to build collaborative teams, empowering them with effective strategies for aligning curriculum, instruction, and assessment to ensure all students receive high-quality mathematics instruction.

Toncheff served as an active board member of NCSM for over eleven years and was president of the organization. In addition, she was a cofounding board member and past president of Arizona Mathematics Leaders. In 2009, she was named Phoenix Union High School District Teacher of the Year. In 2014, she received the Copper Apple Award for leadership in mathematics from the Arizona Association of Teachers of Mathematics, and she received the Arizona Mathematics Leaders Leadership Award in 2022.

Toncheff earned a bachelor of science from Arizona State University and a master of education in educational leadership from Northern Arizona University.

To learn more about Mona Toncheff's work, follow her @Toncheff5 on X.

**Jennifer Deinhart** is an educational consultant and K–8 mathematics specialist. She is currently working as a mathematics instructional coach at Rose Hill Elementary, part of Fairfax County Public Schools. During her time at Mason Crest Elementary in Annandale, Virginia, the school was recognized as the first national Model PLC school to receive the DuFour Award. A passionate educator with more than twenty years of experience working with diverse populations within Title I schools, Deinhart leads collaborative teams of teachers to provide quality mathematics instruction.

As a national presenter, Deinhart guides educators in effective teaming practices, such as the collection of quality evidence of student learning, progress monitoring that leads to intentional targeted re-engagement, and systematic processes for student self-assessment and goal setting. She joined the Mathematics at Work author team for *Mathematics Unit Planning in a PLC at Work: Grades PreK–2* and *Grades 3–5*. She has led multiple districts in adapting pacing guides and developing curriculum resources, and she continues to support schools around the United States in deeply learning and implementing the PLC at Work process.

Deinhart received a bachelor's degree from Buffalo State University, a State University of New York campus, and a master of education degree specializing in K–8 mathematics leadership from George Mason University.

To learn more about Jennifer Deinhart's work, follow her @jenn_deinhart on X.

**Brian Buckhalter, EdD**, is a mathematics education coach and consultant specializing in supporting teachers and students in building conceptual understanding in tandem with procedural application of mathematics skills and concepts, bridging best practices from research to instructional practices in the classroom, guiding and supporting instructional coaches, and developing and implementing effective assessments. He is the founder and lead consultant of Buck Wild About Math, LLC.

Buckhalter previously served as both a middle school teacher and a district mathematics coach for the Oxford School District in Mississippi. He is also a former graduate research fellow for the Center for Mathematics and Science Education at the University of Mississippi, where he helped establish PLCs across several school districts in Mississippi. He is passionate about fostering an understanding of—and love and appreciation for—the role mathematics plays in our daily lives. He has facilitated numerous professional development sessions and delivered keynote addresses and conference sessions across the United States.

Dr. Buck, as he is known in the mathematics education community, has served in several capacities in local, state, and national organizations, including president of the Mississippi Math Specialist Network; NCSM awards chair, director of professional learning, and regional team lead for Mississippi; and editor for NCSM's *Journal of Mathematics Education Leadership*. He is a member of the Mississippi Council of Teachers of Mathematics and NCTM. He also served as a member of the writing team for NCSM's *Essential Action Series: Framework for Leadership in Mathematics Education*.

Buckhalter earned a bachelor of arts degree in elementary education from Dillard University, and both a master's degree and a doctorate in teacher education from the University of Mississippi.

To learn more about Brian Buckhalter's work, follow him @Buckwildabtmath on X.

To book Sarah Schuhl, Mona Toncheff, Jennifer Deinhart, or Brian Buckhalter for professional development, contact pd@SolutionTree.com.

# Foreword

*By Mike Mattos*

To all who read this outstanding book, I sincerely thank you! By seeking out best strategies at Tier 1 and Tier 2, you are acting on a commitment to ensure *all* students develop a deep understanding of rigorous mathematical concepts and procedures. Higher-level thinking skills, such as logical reasoning, problem solving, discovering solution pathways, and validating conclusions are needed to access post-secondary education and career opportunities in the global economy.

When educators serve as gatekeepers to grade- or course-level mathematics expectations, allowing only the students deemed most capable to participate, they are opening—and shutting—doors of lifelong opportunity. For any school or district truly committed to students' future success, ensuring *every* student learns essential mathematic standards requires a deep commitment to the collaborative processes proven to achieve this goal and the courage to discontinue counterproductive practices and assumptions.

Teaching students the mathematical skills and knowledge needed for future success is a cumulative effort from grades preK–12. Every year is a critical piece of the puzzle that collectively creates the portrait of a confident, lifelong learner, with every grade building upon the last and digging deeper into critical concepts and procedures. When students master each grade's essential mathematics standards, they enter the next school year well-prepared. However, when students fail to learn grade-level essential curriculum, it usually leads to future academic struggles, lowered academic expectations, and a fixed mindset of "just not being good" at mathematics. The key to ensuring every student's future mathematics success is this: *Make sure every student learns at grade level or higher every year.* That is the goal of Tier 1 and Tier 2 in a multitiered system of supports (MTSS), and this book provides the proven strategies—the roadmap—needed to make this outcome a reality.

Since 2009, my focus as an educator has been helping schools and districts work collaboratively and create effective systems of interventions. Throughout my travels, I find that most educators learn about the imagery and terminology of the research-based intervention process Response to Intervention (RTI; also known as a multitiered system of supports). This common vocabulary is traditionally captured in a pyramid diagram divided into three tiers, as shown in figure F.1.

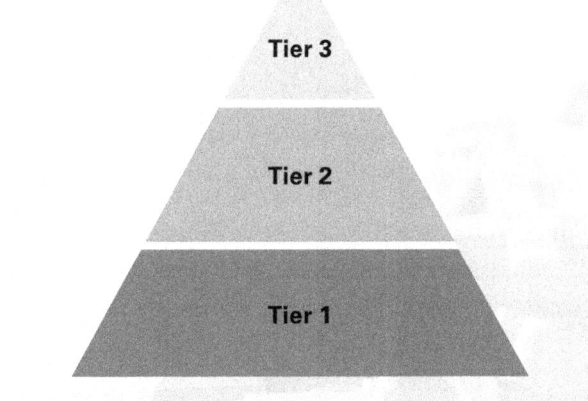

**Figure F.1: Traditional three tiers of RTI/MTSS.**

While this vocabulary is used frequently at most schools, there is inconsistency on how educators interpret the primary purpose of each tier and how the tiers are designed to collectively meet students' diverse needs. Without an accurate, shared understanding of the larger process, it is difficult to fully leverage the Tier 1 and Tier 2 strategies described in this book. To this end, the following is a clear and logical description of the overall process to enhance the description shared by the authors in chapter 1 (pages 11–13).

A multitiered system of supports (MTSS) is designed to meet three essential learning outcomes needed to ensure high levels of learning for all students (Mattos et al., 2025). For the purposes of this book, the focus is mathematics.

If the goal is to ensure every student learns mathematics at grade-level or higher every year, then *all* students must have access to essential grade-level standards as part of their core instruction. Students cannot learn mathematics at grade level if they are learning remedial, below-grade-level curriculum. This is the primary purpose of Tier 1. See figure F.2.

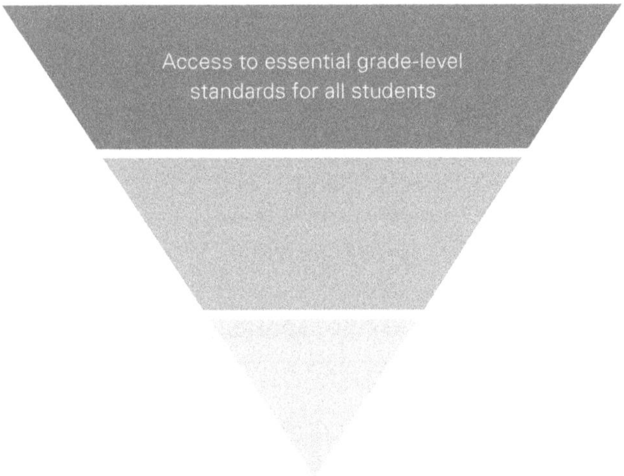

*Source: Mattos et al., 2024, p. 14.*

**Figure F.2: Tier 1—Access to essential grade-level standards for all students.**

Some students will not master every new essential mathematics grade-level standard by the end of a unit of study, but all students must master essential curriculum by the end of the school year to be ready for the next grade or course. That is the primary purpose of Tier 2: to systematically and collectively provide additional time and support to learn essential mathematics grade-level standards. See figure F.3.

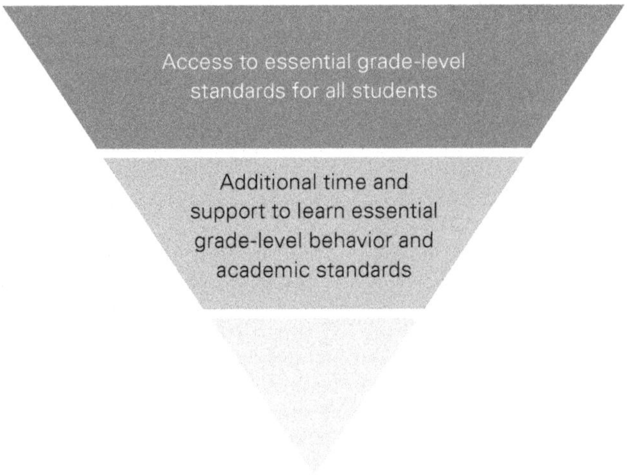

*Source: Mattos et al., 2024, p. 15.*

**Figure F.3: Tier 2—Additional time and support to learn essential grade-level behavior and academic standards.**

Some students will undoubtedly enter a new school year lacking mathematics prerequisite skills they should have mastered in prior years. These students will need intensive reinforcement to support these foundational concepts and procedures. That is the purpose of Tier 3: to provide intensive reinforcements in mathematics (and other) foundational skills from prior years. See figure F.4.

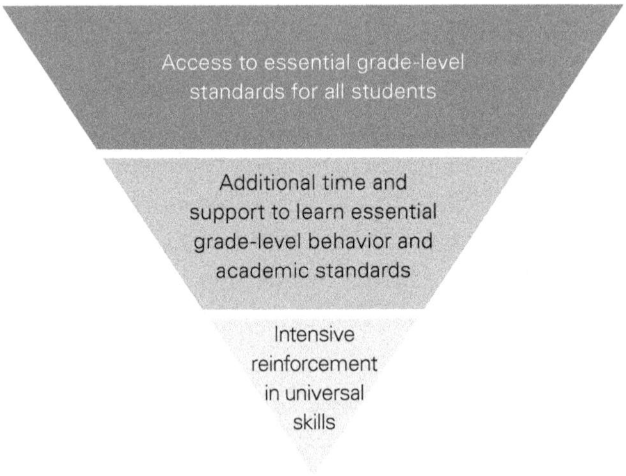

*Source: Mattos et al., 2024, p. 16.*

**Figure F.4—Intensive reinforcement in universal skills.**

And most importantly, some students require all three of these outcomes to learn mathematics at high levels.

1. Access to essential mathematics grade-level standards for all students (Tier 1)
2. Additional time and support to learn essential mathematics grade-level behavior and academic standards (Tier 2)
3. Intensive reinforcement in mathematics foundational skills from prior years (Tier 3)

These three outcomes collectively create a multitiered system of supports to meet the learning needs of all students in mathematics. Please note the term *all*. Students are not targeted or divided by educational labels like *regular education*, *special education*, *Title I*, or *English learner (EL)*. All students receive Tier 1 instruction, and any student can receive Tier 2 and/or Tier 3 supports if they demonstrate the need. This inclusive approach is why the RTI at Work pyramid is inverted (Mattos et al., 2025). Instead of incorrectly viewing the tiers as a pathway to potential special education identification, it is instead an ongoing process to meet the specific learning needs of each student. See figure F.5.

It is impossible to provide all three tiers of support within an individual teacher's classroom. The entire staff needs to work collaboratively and take collective responsibility for student learning. This is best achieved when a school functions as a Professional Learning Community (PLC) at Work, as explained in chapter 1 (page 9).

The pyramid shape of the tiers visually captures the importance of Tier 1 and Tier 2. As teachers get better at collaboratively using effective mathematics teaching strategies at Tier 1, fewer students will need additional help to learn essential grade-level curriculum. This means fewer students will need Tier 2 supports. And as a school or district collectively gets better at effective teaching practices and processes at Tier 2, more students will master new essential mathematics curriculum by the end of the year. So, over time the number of students who will need intensive Tier 3 help diminishes.

Conversely, when Tier 1 practices are ineffective, a school's Tier 2 interventions become overwhelmed with too many needs. This makes Tier 2 less effective, which will likely lead to more students needing Tier 3 help the following year. By committing to best mathematics teaching strategies at Tier 1 and Tier 2, fewer students will need help at the next tier of support. And that is exactly what this book provides: how to leverage the PLC at Work process to implement best mathematics teaching strategies at Tier 1 and Tier 2!

So, I will close this foreword how I started—by thanking you for committing your valuable time to reading this outstanding book! As a fellow educator who started his career teaching middle school mathematics, I found the recommendations in this book to be proven, practical, and doable. I have no doubt that applying what you learn is going to improve your mathematics teaching, increase the collective efficacy of your colleagues, and change the trajectory of your students' lives.

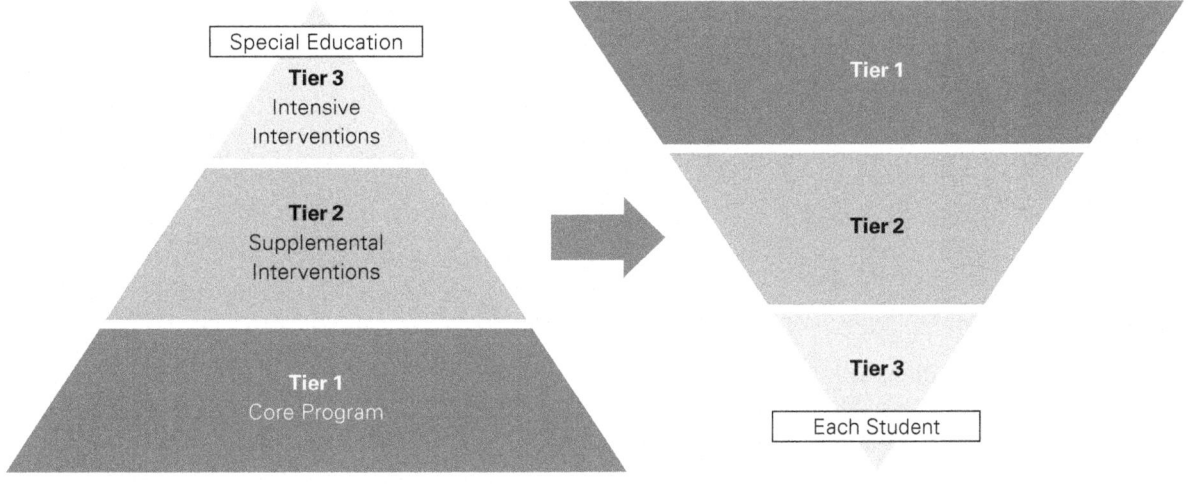

*Source: Mattos et al., 2024, p. 13.*

**Figure F.5: Inverted RTI/MTSS pyramid.**

# Introduction

*Let's be very clear here: The data prior to the pandemic did not reflect an education system that was on the right track. The pandemic simply made it worse . . .*

—Miguel Cardona, U.S. Secretary of Education

When students learn mathematics in grades preK–12, they learn to logically reason and problem solve. They learn to discover solution pathways, justify or validate their conclusions, and discuss mathematics with their peers. Most importantly, students learn to use mathematics to quantify and make sense of their world.

Learning mathematics at grade level and beyond means students have options and opportunities for their futures after graduating high school. And, to graduate with the strong mathematical understanding required, students need to learn grade- or course-level mathematics each school year before moving to the next grade level or course. Unfortunately, too many students think learning mathematics is difficult, especially when there is an early focus on learning basic facts and doing mathematics quickly. Instead of being mathematically literate at grade or course level with reasoning and problem-solving skills, students move to the next grade or course with learning gaps, which means they are not on track to graduate. This can happen as early as grades preK–2. There is a need to work collaboratively as educators through strong core instruction and intervention to ensure every student learns grade- or course-level mathematics or beyond during every year of their preK–12 schooling.

## The Current Reality

In the aftermath of the COVID-19 pandemic, the National Assessment of Educational Progress (NAEP) mathematics assessments show glaring performance differences along racial, ethnic, and socioeconomic lines, as well as by ability status and language background. Fourth-grade mathematics scores decreased in forty-three U.S. states, and eighth-grade mathematics scores declined or had no significant change across all states (Nation's Report Card, n.d.). In fact, the average national score for fourth and eighth grades decreased to scores not seen since 2004 (Institute of Education Sciences, 2022). Unfortunately, U.S schools and districts are actively seeing the need for current student mathematics performance to improve across all vertical grade bands and courses.

Meanwhile, the number of jobs that requires strong number sense and problem-solving skills is increasing. Jon Marcus (2023) of *The Hechinger Report* shares the amount of jobs that "use arithmetic and apply advanced techniques to make calculations, analyze data, and solve problems" will increase by more than 30,000 per year through the end of this decade (p. 3). Students developing strong number sense will better incorporate the mathematics needed in the future job market. Stanford economist Eric A. Hanushek has estimated that, if mathematics declines are not reversed, current preK–12 mathematics students will earn between 2 percent and 9 percent less over their careers, depending on which state they live in, than students educated just before the start of the pandemic (Marcus, 2023).

There are a few other critical issues at play when working to close mathematics learning gaps. It is difficult to reverse mathematics declines when students are not in school. On an average day during the 2022–2023 school year, close to 10 percent of K–12 students in the United States were not present at school. About one-quarter of the students qualified as chronically absent, meaning they missed at least 10 percent of school days, which translates to about three and a half weeks of school missed (Leonhardt, 2023). This level of absenteeism exceeded pre-COVID-19 numbers (Leonhardt, 2023). In addition to absenteeism, it is difficult to reverse mathematics declines without highly qualified mathematics teachers. Unfortunately, the pandemic and the negative public view of education in the pandemic's aftermath exacerbated an already growing mathematics teacher shortage (Klein, 2022).

Teachers and collaborative teams face the current challenge of accelerating the learning of students with significant needs in mathematics to grade or course level and beyond. Yet, the reality is that students not being ready for grade-level mathematics learning is nothing new. All too often, students were not ready to learn new content and skills when they started school each year. The variances in student learning are larger than before the pandemic.

Through our work over the last few years with mathematics teachers and teacher teams in professional development experiences, we noticed they are frustrated at the wide learning variances in students entering their classrooms. Once teachers have tried everything in their instructional tool kit, their general response is to engage in remediation by teaching students below grade level, focusing on basic facts and algorithms, sending students to an interventionist, or creating an intervention that is not targeted and specific but tries to reteach several years' worth of mathematics at once. These remediation practices are unproductive. Instead, there is a need for acceleration—growing student learning along learning progressions by teaching grade- or course-level essential standards and prior-knowledge skills for those essential standards (Kramer & Schuhl, 2023).

## Mathematics in a PLC at Work

Accelerating every student's learning requires teachers to continue learning (Barth, 1990). Reporter and educational writer Karin Chenoweth (2021) explains that teachers need to learn from one another and collectively work toward helping all students learn. She writes:

> It is long past time to acknowledge that it is impossible for individual educators to know all there is to know about making kids smarter. There is simply too much to know. It is only by pooling their knowledge and learning from expertise that educators can possibly expect to help all kids. (p. 137)

Teachers learn from one another and collectively take ownership for the learning of all students across their grade, course, or department when they work in collaborative teams within a Professional Learning Community at Work®. Collaborative mathematics teams are foundational to the Mathematics in a PLC at Work framework and process. The three big ideas of a PLC at Work include the following.

1. A focus on learning
2. A collaborative culture and collective responsibility
3. A results orientation (DuFour et al., 2024)

Within these three big ideas, teacher teams answer the following four critical questions for a PLC at Work established by educational researcher Richard DuFour and colleagues (2024):

> 1. What knowledge, skills, and dispositions should every student acquire as a result of this unit, this course, or this grade level?
> 2. How will we know when each student has acquired the essential knowledge and skills?
> 3. How will we respond when some students do not learn?
> 4. How will we extend the learning for students who are already proficient? (p. 44)

You and your mathematics colleagues learn together when you collaboratively make sense of essential standards, create common assessments, analyze student work to identify effective instructional practices, look for trends in student thinking, and create a team response to student learning. In turn, students learn, too.

Not all students learn the same way or at the same rate. Sarah Schuhl and colleagues (2024) state:

> When students have not yet learned essential learning standards, the collaborative teacher team response in a PLC culture is not to stop and reteach all students. Such an action would be done at the expense of students learning the remaining essential standards that are part of the guaranteed and viable curriculum for your grade level or mathematics course. (p.121)

In the PLC at Work process, your team creates a collective response to student learning after gathering information from common mid-unit and end-of-unit assessments. The information gathered provides insight into errors, misconceptions, or mistakes students are making so you can create a collective response to re-engage students in Tier 1 (core instruction) or Tier 2 (supplemental intervention and extension) instruction. See chapter 1 (page 9) for information related to Tier 1 and Tier 2 instruction.

The Mathematics in a PLC at Work framework in figure I.1 shows the collaborative team actions needed when teachers are learning together and answering the four critical questions.

As shown in figure I.1, you and your team implement effective mathematics instruction and intervention practices across collaborative teams with a collective responsibility for student learning.

And, as noted in the Mathematics at Work *Every Student Can Learn Mathematics* series, teacher teams make sense of the mathematics that students will learn

| *Every Student Can Learn Mathematics* series' Team Actions Serving the Four Critical Questions of a PLC at Work | 1. What knowledge, skills, and dispositions should every student acquire as a result of this unit, this course, or this grade level? | 2. How will we know when each student has acquired the essential knowledge and skills? | 3. How will we respond when some students do not learn? | 4. How will we extend the learning for students who are already proficient? |
|---|---|---|---|---|
| *Mathematics Assessment and Intervention in a PLC at Work, Second Edition* | | | | |
| **Team action 1:** Develop high-quality common assessments for the agreed-on essential learning standards. | ■ | ■ | | |
| **Team action 2:** Analyze and use common assessments for formative student learning and intervention. | | | ■ | ■ |
| *Mathematics Instruction and Tasks in a PLC at Work, Second Edition* | | | | |
| **Team action 3:** Develop high-quality mathematics lessons for daily instruction. | ■ | ■ | | |
| **Team action 4:** Analyze and use effective lesson design elements to provide formative feedback and build student perseverance. | | | ■ | ■ |
| *Mathematics Homework and Grading in a PLC at Work* | | | | |
| **Team action 5:** Develop and use high-quality common grading components and formative grading routines. | ■ | ■ | ■ | ■ |

*Source: Schuhl et al., 2024, p. 3.*

**Figure I.1: Mathematics in a PLC at Work framework.**

in each unit, create common assessments to gather evidence of student learning by target, create time in the day or unit for Tier 1 instruction and Tier 2 interventions, practice the routines and foundations of effective lesson design, and establish grading structures that align with student learning.

The National Council of Teachers of Mathematics ([NCTM], 2000) recommends that students have an opportunity to develop an understanding of mathematical concepts and procedures by engaging in meaningful instruction. *Conceptual knowledge* means developing a deeper understanding of mathematics concepts by connecting new concepts and skills to those previously learned and understanding the relationships and patterns among these different pieces of information and the contexts in which they are used (Miller & Hudson, 2007). All students learn mathematics when they conceptually understand and develop accurate and efficient procedures (including standard algorithms) to solve computational problems (U.S. Department of Education & National Mathematics Advisory Panel, 2008).

How do you ensure all students are learning grade-level mathematics content? Are you looking for additional strategies to ensure students are learning grade- or course-level content? Then you have come to the right place!

## What's in This Book

This book shares mathematics strategies for teachers and collaborative teams to use with students in Tier 1 and Tier 2 learning experiences. These strategies are not necessarily assigned to Tier 1 or Tier 2 instruction, specifically; teachers can often effectively use them in both tiers to grow student learning. However, when appropriate, we share distinctions for how to use the strategies in Tier 1 or Tier 2. Note that throughout the book we refer to primary students, intermediate students, and middle and high school students. Generally, these translate to grades preK–2 (primary), grades 3–5 (intermediate), grades 6–8 (middle school), and grades 9–12 (high school or secondary).

The strategies in this book build on the work and research from NCTM, National Council of Supervisors of Mathematics (NCSM), Association of State Supervisors of Mathematics (ASSM), Robert J. Marzano, John Hattie, and Solution Tree Press's *Every Student Can Learn Mathematics* series, among others. Through the three parts in this book, you will gain ready-to-implement strategies, regardless of your school or district's curriculum resources. You and your collaborative team will learn how to collectively respond to student learning and ensure your students understand mathematics at grade or course level and beyond.

### Part 1: A Culture of Learning

Part 1 includes chapters 1–2. Chapter 1 explores the role of collaborative teams in a Response to Intervention (RTI) at Work™ structure or multitiered system of supports. You and your collaborative team explore how to ensure students learn grade- or course-level standards through intentional Tier 1 instruction and assessment and Tier 2 interventions and extensions offered during the school day. The chapter also discusses how to teach mathematics with a focus on conceptual understanding and how to intentionally design lessons using a Tier 2 intervention planning tool.

Chapter 2 guides you to create a classroom culture in which students see themselves as a valuable part of a community of learners. This includes considering physical space and strategies for sharing expectations with students to build a strong sense of belonging in Tier 1 and Tier 2 learning experiences.

### Part 2: Mathematics Foundations

Part 2 includes chapters 3–7. Chapter 3 explores strategies and routines to support students learning grade- or course-level content. To learn grade- or course-level standards, students should be actively engaged in learning grade- or course-level mathematics every day during Tier 1 instruction. You and your team scaffold or differentiate that learning using high-level-cognitive-demand tasks or grow it through strategically chosen leveled tasks in Tier 2 learning experiences.

Chapter 4 emphasizes the importance of making connections to prior knowledge and provides strategies for doing so intentionally. It describes how to help students understand how concepts evolve through the grades and how to view mathematics standards as a coherent progression, with each previous learning experience supporting current learning.

Chapter 5 focuses on developing number sense and explores strategies, such as the use of tools, pictorial models, and verbal and written expressions. You and your team examine how to provide students with numerous opportunities to engage in number routines in the early grades, such as subitizing and counting. Your team designs targeted Tier 2 learning experiences to build strong numeracy foundations when students need extra time and support to master and apply the number sense needed to learn grade- or course-level standards. A strong sense of numbers develops the flexibility needed to access grade- and course-level mathematics.

Developing problem-solving reasoning and applying it to concepts and skills is at the heart of mathematics. Chapter 6 explores strategies, tools, and models for approaching problems in various ways and offers different methods students can use to communicate their strategies and reasoning with one another.

Chapter 7 specifies strategies to develop procedural fluency. Students in Tier 1 need exposure to multiple strategies involving manipulatives, models and pictures, and numbers, along with time to understand which strategies work most effectively and efficiently in different situations. Students still developing procedural fluency in Tier 2 may need explicit guided instruction around a limited set of strategies to re-engage in learning through concrete models, pictures, numbers, and mathematical symbols.

### Part 3: Student Engagement

Part 3 includes chapters 8–11. Chapter 8 discusses the importance of mathematical language for both teachers and students. Using precise mathematical language throughout your lessons strengthens students' understanding. Students need to develop a robust language that includes mathematics vocabulary and notations to communicate precise mathematical thinking in Tier 1 and Tier 2.

Chapter 9 examines how to orchestrate opportunities for students to engage in purposeful discourse with their peers to support an understanding of mathematics. You and your team will explore structures, tools, and strategies to facilitate meaningful, student-generated discourse.

Chapter 10 explores how to generate feedback for student learning. Meaningful feedback drives continuous learning in the classroom and is provided by the teacher or fellow students. This chapter investigates how to continuously utilize descriptive feedback as a learning strategy for students in Tier 1 and Tier 2 learning experiences.

Finally, chapter 11 explores how to build student investment into your daily instruction and assessment routines. This chapter reviews strategies to help students set goals and monitor their progress toward specific learning targets and helps you determine how preK–12 students can articulate what they have and have not yet learned to frame their Tier 1 and Tier 2 instruction.

Each chapter shares the *why* behind the chapter topic and clarifies and provides strategies for how to grow student learning in Tier 1 and Tier 2. The end of each chapter features questions you and your team can use as a reflection and planning tool to identify your next steps.

The book concludes with three appendices. Appendix A offers a series of data analysis protocols you and your team can reference, and appendix B provides a cognitive-demand-level task analysis guide. Finally, appendix C offers a list of team actions to avoid and consider for Tier 1 and Tier 2.

Figure I.2 (page 6) describes the connections between the three parts of the book and how to support your intentional planning of Tier 1 instruction and Tier 2 interventions.

## Conclusion

You and your collaborative team make decisions every day that impact students' mathematics learning. You determine the specific content students need to learn unit by unit, including both the rigor of the standards and the types of tasks they must complete to demonstrate their learning. By answering the four critical PLC at Work questions, your collaborative team can focus on the instruction, assessment, intervention, and extensions needed to ensure students learn mathematics at grade or course level and beyond. Together, the Tier 1 and Tier 2 learning experiences students receive impact their successful understanding and application of mathematics for years to come.

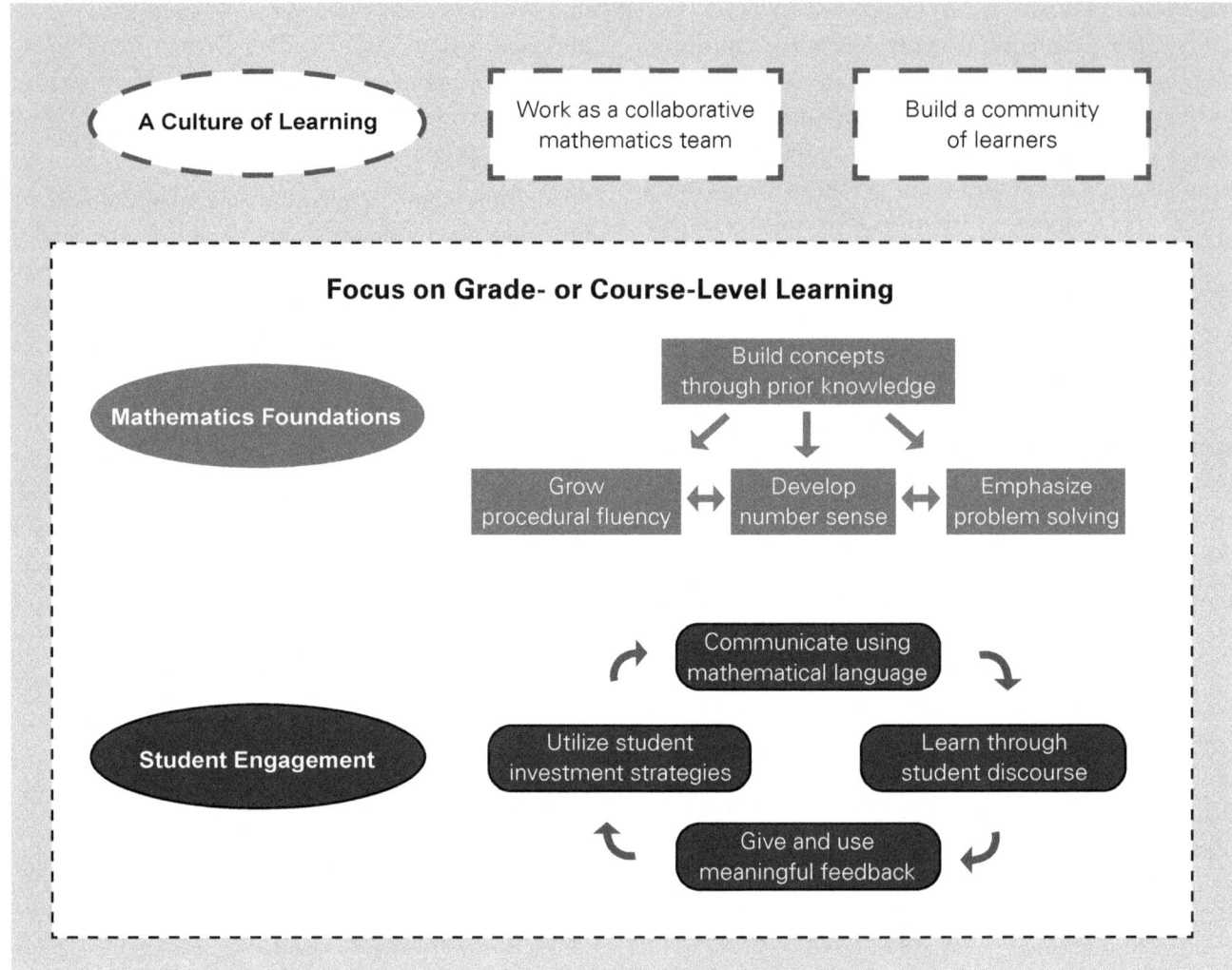

**Figure I.2: Tier 1 and Tier 2 mathematics instruction in a PLC at Work.**

*Visit **go.SolutionTree.com/MathematicsatWork** for a free reproducible version of this figure.*

# PART 1

# A Culture of Learning

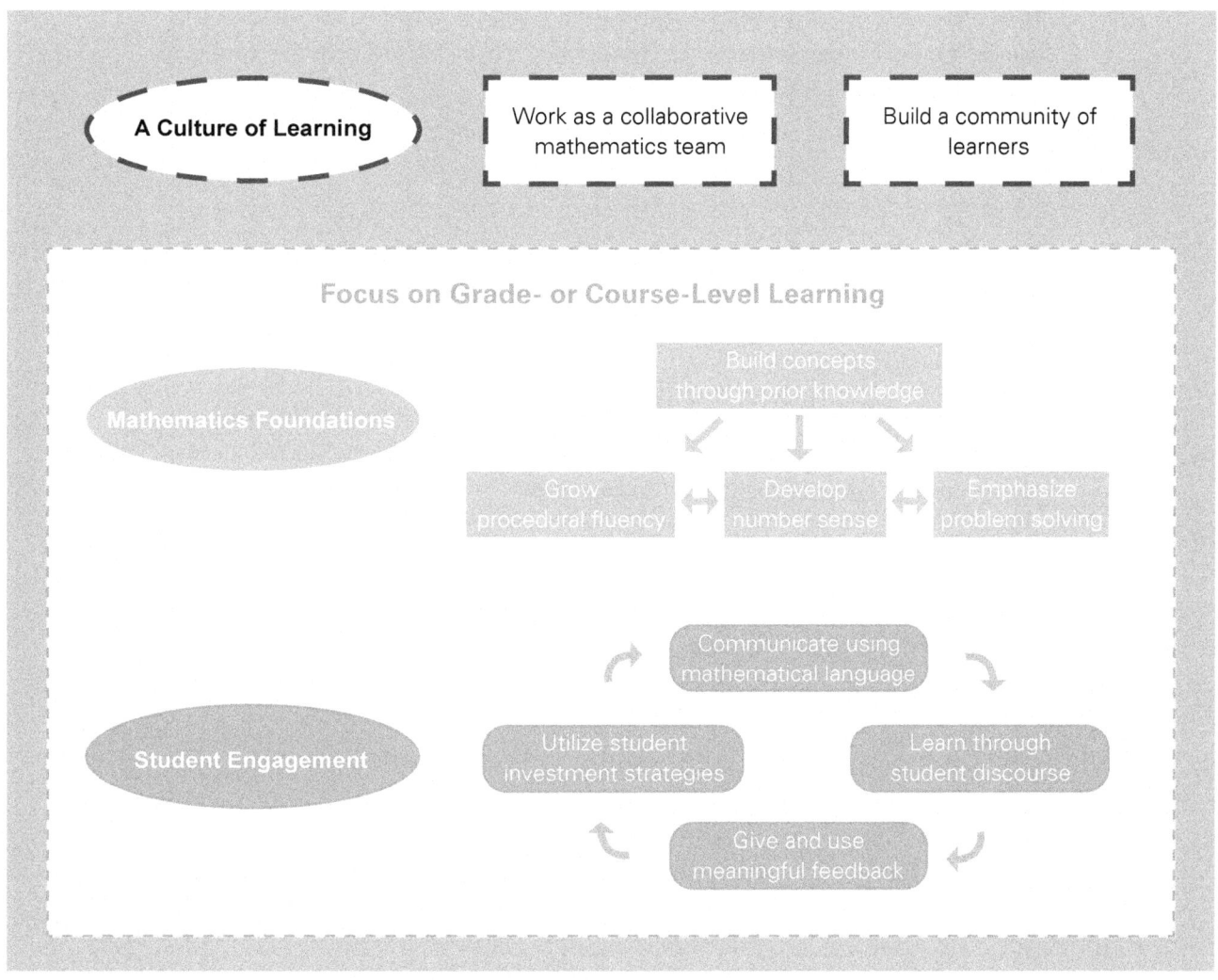

> Early in Sarah's teaching career, she taught algebra 1, the entry-level course at her high school. There were several students who clearly demonstrated their learning of preK–8 mathematics skills and were ready to learn algebra 1, and there were others who struggled from the first day to access algebra 1 content because they had moved from one grade to the next without learning grade-level mathematics. Sarah felt tremendous pressure when she was tasked with ensuring all students learned algebra 1 content.
>
> Though she worked with other teachers at her high school, they did not yet function as a collaborative team in a PLC at Work, so Sarah was left to determine on her own how to address learning gaps with no additional time in the school day for interventions. She tried to reteach many mathematics topics before addressing algebra 1 content but found that strategy bored some students and left others still confused. Reteaching a bunch of content in a short amount of time did not seem to help students gain the learning needed to be successful in algebra 1, and it also left less time to learn the current-year algebra 1 standards. Neither Sarah nor her students were set up for success.

# CHAPTER 1

# Work as a Collaborative Mathematics Team Focused on Student Learning

*If not us, then who?*
*If not now, then when?*

—John Lewis

In many classrooms, including Sarah's (see story on page 8), students have large variances in mathematics learning and understanding. While Sarah's intention to help all students learn was admirable, her execution fell short. Student learning inequities occur across the team and school when each teacher decides individually what to reteach and teach. Beginning the year by reviewing everything students should have learned the year before is not productive. Some students are bored, while others are frustrated. Instead, to accelerate student learning, work with your collaborative team to determine essential learning standards and systematically provide targeted intervention (and extension) supports to students through Tier 1 and Tier 2 learning experiences. The latter approach will grow student confidence in their ability to learn mathematics.

Imagine a world in which students learn, laugh in joy, and curiously explore mathematics together. Imagine students discovering they can use mathematics to quantify their world and pursue deepening their own mathematical reasoning to apply to other courses and life experiences. Such a scenario asks students to be problem solvers with a belief that they can do mathematics. It requires them to persevere and collaborate in their learning. Students think about reasonable answers and try solution pathways to justify their thinking.

Use the rubric in figure 1.1 (page 10) as you and your team reflect on your current Tier 1 and Tier 2 instructional practices and determine the next steps to ensure every student learns grade- or course-level mathematics. We address each criterion in the rubric throughout the rest of this book.

Mathematics teaching and learning are complex. For too many students, mathematics becomes a gatekeeper to desired opportunities during and after their preK–12 experiences. There are myriad reasons that contribute to mathematics learning for each student and, therefore, myriad reasons a student has not yet learned grade-level mathematics. Regardless, there are actions a teacher or collaborative team can take to accelerate every student's learning to grade level and beyond in mathematics. These actions require strong instructional strategies used in core instruction and during any intervention or extension re-engagement opportunities.

## The *Why* and the *What* of Working as a Collaborative Mathematics Team Focused on Student Learning

There are many district, school, or classroom responses when teachers realize not all students are grade- or course-level ready in mathematics. A few common strategies that research shows are *not* impactful include the following.

- Ability grouping/tracking students for core instruction (Visible Learning Meta$^X$, 2023)
- Retention (Visible Learning Meta$^X$, 2023)
- Remediation (NCTM, NCSM, & ASSM, 2021)
- Tracking teachers; for example, assigning all new high school teachers to teach algebra 1 while veteran teachers teach the upper-level courses (NCTM et al., 2021)

| High-Quality Tier 1 and Tier 2 Mathematics Strategies | Description of Level 1 | Requirements of the Indicator Are Not Present | Limited Requirements of the Indicator Are Present | Substantially Meets the Requirements of the Indicator | Fully Achieves the Requirements of the Indicator | Description of Level 4 |
|---|---|---|---|---|---|---|
| 1. High-Quality Tier 1 Instruction | • Teachers individually plan each mathematics unit. Instruction covers many standards, and teachers often determine lessons using a textbook or outside resource.<br>• Each teacher responds to Tier 1 student learning themselves. Some may continue teaching without planning for differentiation or intervention during best, first core instruction. | 1 | 2 | 3 | 4 | • Teacher team collaboratively plans each mathematics unit. Instruction is focused on essential learning standards.<br>• Teacher team plans for learning of grade-level essential standards, differentiation, small groups, and time to respond to student learning collectively using data from common mid-unit assessments. |
| 2. A Culture of Learning | • Teachers assume students see themselves as part of the learning community. Some students work well together, and others do not. Teachers expect students to behave, interact, and learn from one another but may not intentionally teach students those routines. | 1 | 2 | 3 | 4 | • Teacher team implements strategies so students see themselves as valuable contributors to a community of learners.<br>• Teacher team discusses how to clarify behavior and academic expectations and considers how to arrange the classroom to promote students learning from one another. |
| 3. Mathematics Foundations | • Teachers may teach students below grade level and address all prior knowledge from many grades.<br>• Teachers focus instruction on procedural fluency without conceptual understanding, and problem solving is often done with the teacher leading the discussion. | 1 | 2 | 3 | 4 | • Teacher team plans for grade-level learning of essential standards and identifies the connected prior knowledge to address.<br>• Teacher team also develops student number sense, conceptual understanding, procedural fluency, and student problem-solving skills. |
| 4. Student Engagement | • Teachers have inconsistent routines and expectations.<br>• Teachers each determine the mathematical language students need to communicate and vary in how the language is taught, if at all.<br>• Students may not know how to learn through discourse with peers.<br>• Teachers have different student reflection and goal-setting tools or do not yet have such tools. | 1 | 2 | 3 | 4 | • Teacher team identifies effective instructional routines.<br>• Teacher team teaches the mathematical language students need for communicating.<br>• Teacher team ensures students learn from one another through meaningful discourse.<br>• Teacher team creates reflection and goal-setting tools for students to track and reflect on their learning. |
| 5. High-Quality Tier 2 Intervention | • Additional time is not clearly allocated for teachers across the team to re-engage students in learning. Intervention varies by teacher. Some teachers may only use progress-monitoring data (not common assessments) to create Tier 2 student groups.<br>• Teachers may send students to an interventionist or only use computer programs with minimal to no live instruction as an intervention approach.<br>• Teachers design interventions and extensions for their own students. | 1 | 2 | 3 | 4 | • Additional time is clearly allocated for teacher teams to re-engage their students in learning essential standards or prior knowledge standards.<br>• Teacher team uses data from common assessments to frequently and flexibly create targeted interventions and extensions and shuffle students, as needed, for that learning.<br>• Tier 2 involves some form of live instruction.<br>• Teacher team monitors the effectiveness of its Tier 2 student groups.<br>• Students monitor their learning growth. |

**Figure 1.1: High-quality Tier 1 and Tier 2 mathematics strategies rubric.**

*Visit go.SolutionTree.com/MathematicsatWork for a free reproducible version of this figure.*

Fortunately, there are also research-affirmed strategies that *do* work. In *Catalyzing Change in Early Childhood and Elementary Mathematics: Initiating Critical Conversations*, NCTM (2020) states every student should be in a heterogeneous class for learning core instruction. It is recommended that interventions should focus on prior knowledge that is connected to the standard being taught as part of a learning progression (NCTM et al., 2021). In other words, students need to learn grade- or course-level standards to be ready for the school year, and their learning is accelerated through strategic interventions.

In *Principles to Actions: Ensuring Mathematical Success for All*, NCTM (2014) uses a meta-analysis of studies to conclude that teachers have productive and unproductive beliefs. Teachers with unproductive beliefs teach with a focus on algorithms and basic facts. This works for some students, but not all. Teachers with productive beliefs teach students to reason and connect concepts, apply learning, and work toward procedural fluency. They choose tasks to develop student perseverance and create a safe classroom community in which students try new ideas, struggle productively, and learn through their successes and mistakes. Teachers with productive beliefs ensure more students learn using research-affirmed strategies designed for all students.

Educational researcher John Hattie compiled a list of influences that positively or negatively impact student learning using 2,103 meta-analyses of over 132,000 studies involving over three million students (Visible Learning Meta$^X$, 2023). In his studies, an effect size of 0.4 is the learning expected of a student who ages one year and has an average teacher. Influences above 0.4 have the possibility of accelerating student learning. As of 2024, the following are the two largest influences on student learning.

1. Collective teacher efficacy (effect size 1.34)
2. Teacher estimates of student achievement (effect size 1.29; Visible Learning Meta$^X$, 2023)

*Collective teacher efficacy* means teachers work together to grow practices that ensure every student learns. Teachers collectively believe they can move student learning to grade level and beyond. Teacher estimates of student achievement refer to an educator teaching each student to a level they deem the student can learn. If that level is grade level and beyond, the student learns at grade level or beyond. If that level is below grade level, the student learns below grade level. In other words, collective teacher efficacy asks, "Do we believe in ourselves?" and teacher estimates of student achievement asks, "Do we believe in the students?"

As shared in the introduction (page 1), responding to students who need accelerated learning to grade or course level is larger than any one teacher can accomplish and requires teachers to continuously collaborate and learn together. Engaging in the PLC at Work process means teachers take collective responsibility for student learning and work together to ensure learning across their grade level or course and in their school.

In the Mathematics in a PLC at Work framework shared in the *Every Student Can Learn Mathematics* series, mathematics teams collaboratively answer the PLC at Work four critical questions with a focus on creating and using unit plans (including the meaning of grade- or course-level standards), common assessments, instruction and lesson design routines, team interventions, and team grading practices. Teacher teams answering PLC at Work critical questions 3 and 4 address team interventions and extensions using data and student work collected from common assessments.

Mathematics teams are part of a schoolwide response to intervention (RTI), also called a multitiered system of support. In *Taking Action: A Handbook for RTI at Work*, educator and consultant Mike Mattos and colleagues (2025) present and outline the RTI at Work pyramid, which defines the roles of a school's guiding coalition, collaborative teams, and an intervention team in a systematic approach to ensure students learn.

The purpose of the pyramid is not to identify students for special education but instead, serve students' needs through intentional Tier 1, Tier 2, and Tier 3 learning experiences, interventions, and extensions. For the purposes of this book, we focus on the part of the pyramid referencing teacher team responsibilities, the upper-right portion of the RTI at Work pyramid, shown in figure 1.2 (page 12).

### Tier 1 Preventions

Tier 1 is the prevention level—the goal is to teach mathematics during core instruction. Students learn grade-level mathematics through differentiation, small-group learning, connections to prior knowledge, and

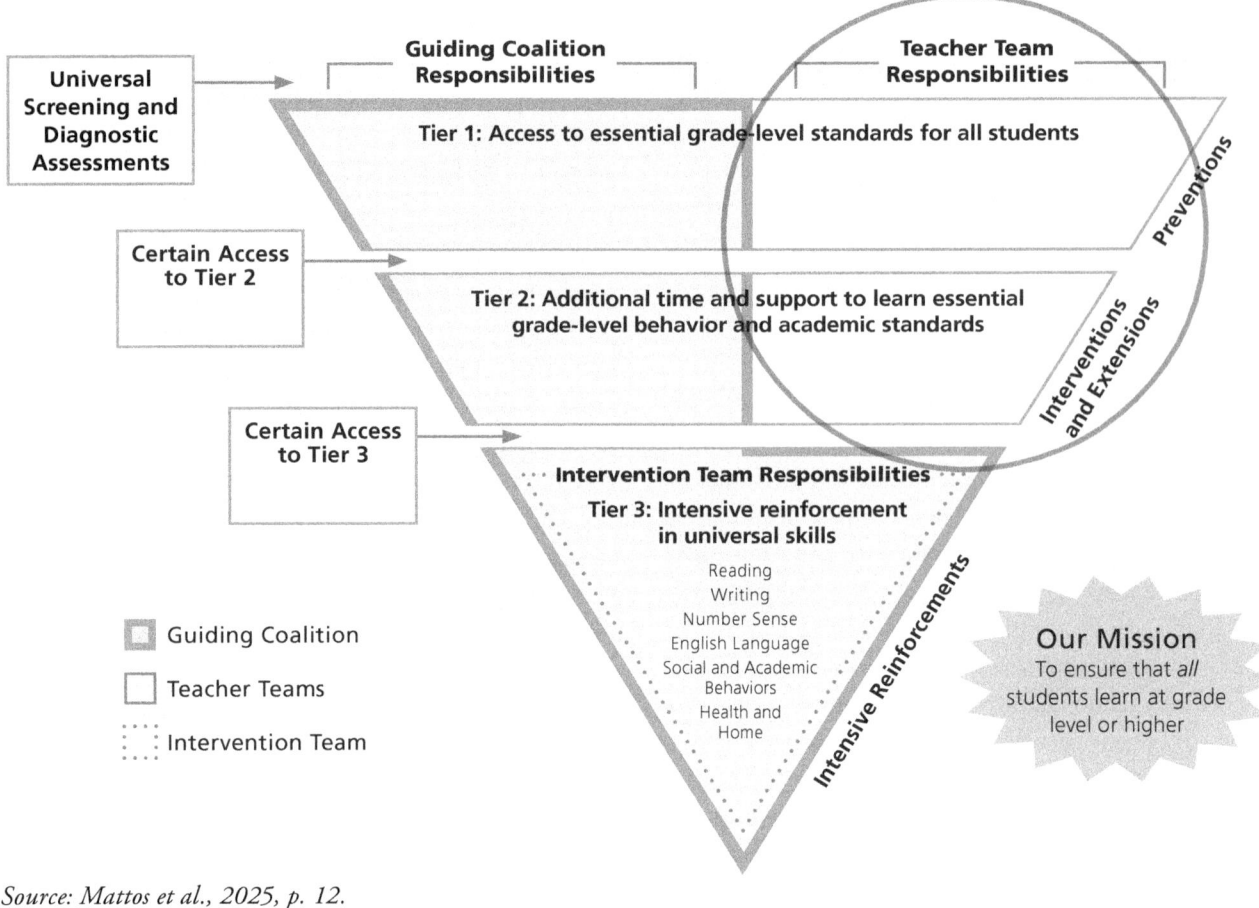

*Source: Mattos et al., 2025, p. 12.*

**Figure 1.2: RTI at Work pyramid.**

use of mathematical tasks incorporating student discourse. As such, students may not need Tier 2 supplemental interventions. Sometimes, Tier 1 core instruction is also called *best first instruction*. Tier 1 instruction includes your team's common mid-unit assessments and collective team response to student learning using, for example, a flex day (day preplanned for intervention), common warm-ups for a week, or small-group learning.

The guiding coalition in Tier 1 (top-left part of the RTI at Work pyramid) ensures collaborative mathematics teams have time to meet each week for planning and that Tier 1 instruction is uninterrupted, among other actions (see *Taking Action* [Mattos et al., 2025] for more information). Uninterrupted Tier 1 instruction means students are not pulled from class for special services when essential standards are being taught, and teachers can count on having their full period or block of time to teach students.

### Tier 2 Interventions and Extensions

Tier 2 is the interventions and extensions level. Mattos and colleagues (2025) recommend scheduling time for Tier 2 outside of core instruction within the school day for at least thirty minutes twice per week. Teams use common assessment data to learn about instructional practices and sort students by targeted need for interventions or extensions (see appendix A, page 173, for data protocols). For the purposes of this book, we focus on interventions in Tier 2. While extensions are also critical, the main goal of this book is to accelerate student learning to grade level through instructional strategies used in Tier 1 and Tier 2 learning experiences when students are not yet learning grade-level mathematics.

Tier 2 interventions focus on grade-level essential standards or prior knowledge standards leading to the current standard being taught and assessed. Teams use common end-of-unit assessment data and student work to inform Tier 2 learning. Time in Tier 2 is not about reteaching the same way students originally learned, nor is it about sitting students in front of a computer program that determines interventions. Tier 2 interventions also do not include sending students out to another person to be retaught; rather, a team owns the Tier 2 student learning needs (and may use additional personnel). Tier 2 learning requires live, targeted, and intentional instruction using strategies like those found in this book. The success of Tier 2 is measured by the number of students who re-engage in learning and demonstrate their new understanding.

The school's guiding coalition ensures there is dedicated time during the school day for Tier 2 interventions, along with the core mathematics instruction provided in Tier 1. The guiding coalition creates systematic intervention and extension time that is fluid and flexible. Students can move in and out as needed during the school day, and intervention support is staffed effectively to address just-in-time needs (connected to what students will learn, are learning, or have finished learning).

### Tier 3 Intensive Reinforcements

Tier 3 is designed to provide intensive reinforcements. A school intervention team is responsible for determining students who, in addition to Tier 1 prevention and Tier 2 intervention, also need intensive reinforcement in foundational universal skills, such as reading, writing, number sense, English language, social and academic behaviors, and health and home (see bottom of the RTI at Work pyramid in figure 1.2). For example, in mathematics, Tier 3 might be used for students with dyscalculia or those needing mathematics strategies to develop number sense and an understanding of operations well below grade level, to name a few. Often, Tier 3 learning happens during the school day with a qualified specialist in addition to learning with you and your collaborative team in Tiers 1 and 2.

In Hattie's research, a systematic and targeted response to intervention, as shown in the RTI at Work pyramid (see figure 1.2), has an effect size of 0.73 and has the "potential to considerably accelerate learning" (Visible Learning Meta$^X$, 2023). Note that it is important not to label students as *Tier 1*, *Tier 2*, or *Tier 3 students*. There are no such students, only students who need Tier 1, Tier 2, or Tier 3 supports.

The work of collaborative teams answering the PLC at Work four critical questions in an RTI at Work focuses on teaching and learning in both Tier 1 and Tier 2. Later chapters address mathematics strategies to use during Tier 1 and Tier 2 instruction. Table 1.1 (page 14) clarifies the work of collaborative teams when addressing Tier 1 and Tier 2 mathematics learning in an RTI at Work. For additional examples and non-examples of team actions, see appendix C (page 181).

To accelerate mathematics learning to grade or course level and beyond, it is essential for collaborative teams to create a collective and targeted response when students have not yet learned. Together, you and your collaborative team plan for student learning during core instruction and determine strategies to use in Tier 2 when students have not learned *yet*—the focus of this book.

## The *How* of Working as a Collaborative Mathematics Team Focused on Student Learning

In the RTI at Work process, collaborative teams answer the four critical questions of a PLC at Work to strengthen core instruction (Tier 1 in the RTI at Work pyramid) and develop a robust and targeted team intervention (and extension) plan, when needed, to ensure every student learns to grade or course level and beyond. Sometimes the intervention occurs in Tier 1 best first instruction, and other times, it occurs during an alternate time of day designated as Tier 2 intervention time. Consider the following actions to focus your Tier 1 and Tier 2 instruction.

### Time in the School Day for Tier 2 Interventions and Extensions

To generate additional time in the school day for Tier 2 interventions, your school might start by using a modified schedule twice per week (such as that used on days when there are assemblies) or taking a few minutes from class periods or instructional blocks to create a thirty-minute block of time for interventions.

If, however, your school does not have additional time built into the schedule for Tier 2 interventions and will not have it anytime soon, you may have to carve out some time during a common time of day when you and your team are teaching mathematics to address Tier 2 interventions. For example, on Tuesdays and Thursdays, you and your team might shorten your Tier 1 mathematics lesson by thirty minutes to use that time for Tier 2 interventions. Secondary schools can also shorten their lessons for the same purpose twice per

Table 1.1: Work of Collaborative Mathematics Teams Focused on Student Learning in Tier 1 and Tier 2

| Work of Collaborative Mathematics Teams Focused on Student Learning in Tier 1 and Tier 2 | |
|---|---|
| <ul><li>Identify and build a shared understanding of the essential mathematics standards in each unit.</li><li>Clarify the prior knowledge standards students need to learn as a connection to Tier 1 or for targeted Tier 2 interventions.</li><li>Implement high-quality lesson design routines during instruction.</li></ul> | |
| **Work of Collaborative Mathematics Teams Focused on Student Learning in Tier 1** | **Work of Collaborative Mathematics Teams Focused on Student Learning in Tier 2** |
| <ul><li>Clarify strategies and models to use when teaching each unit's standards.</li><li>Create and give high-quality common mid-unit and end-of-unit assessments.</li><li>Analyze students' work and data from common assessments to evaluate instructional practices and determine targeted and specific interventions.</li><li>Create and implement team-targeted and specific interventions and extensions as a response to a common mid-unit assessment.</li></ul> | <ul><li>Use common assessment data and student work (often from end-of-unit assessments) to determine students who need Tier 2 interventions.</li><li>Determine instructional strategies and models to use for each targeted intervention or extension.</li><li>Create an assessment to determine whether students learned during Tier 2 experiences.</li><li>Analyze learning data from Tier 2 interventions and extensions to determine next steps for student learning as a team.</li></ul> |

*Visit **go.SolutionTree.com/MathematicsatWork** for a free reproducible version of this table.*

week. Teams might also plan to utilize a flex day, which is a preplanned day built into your team unit calendar to re-engage students in learning after a common mid-unit assessment (Schuhl et al., 2020). Your team can check on the effectiveness of your Tier 2 intervention plan using the Teacher Team Rubric: Tier 2 Mathematics Intervention Program document in *Mathematics Assessment and Intervention in a PLC at Work, Second Edition* (Schuhl et al., 2024). You can visit **go.SolutionTree.com/MathematicsatWork** to download a free reproducible of this rubric.

Figure 1.3 shows options for how your team might schedule time during the school day for Tier 1 core instruction and Tier 2 interventions. While option A in figure 1.3 is ideal—or a similar structure with two days for Tier 2 intervention—you may need to be creative.

Collaborative mathematics teams focus collectively on Tier 1 and Tier 2 learning experiences for all students, leaving Tier 3 (ideally only needed by a few students) to additional experts. Throughout this book, your collaborative team will find instructional strategies to use during Tier 1 instruction and Tier 2 interventions to accelerate students' mathematics learning to grade level and beyond.

## Tier 1 and Tier 2 Instruction on Essential Learning Standards

When answering the first PLC at Work critical question—"What knowledge, skills, and dispositions should every student acquire as a result of this unit, this course, or this grade level?" (DuFour et al., 2024, p. 44)—collaborative teams first determine which standards are essential for students to learn before entering the next grade or course (DuFour et al., 2024; Mattos et al., 2025; Schuhl et al., 2020, 2024). These essential learning standards represent the minimum content that students will learn throughout the year. In addition to teaching the essential standards, you and your team will also teach important-to-know standards (some of the remaining supporting standards for the grade level), but not necessarily any nice-to-know standards.

Essential learning standards are the emphasis for instruction during the unit and common assessments and drive intervention and extension work across a team. Teams accelerate learning of essential learning standards along a learning progression to ensure every student learns at grade or course level or above. Students learning essential standards to grade level or above means they enter the next grade level with critical learning in place. For example, a third-grade

| Possible Schedule | Monday | Tuesday | Wednesday | Thursday | Friday |
|---|---|---|---|---|---|
| A | Tier 1 Instruction / Tier 2 Intervention | Tier 1 Instruction / Tier 2 Intervention | Tier 1 Instruction / Tier 2 Intervention | Tier 1 Instruction / Tier 2 Intervention | Tier 1 Instruction / Tier 2 Intervention |
| | | | Or | | |
| B | Tier 1 Instruction | Tier 1 Instruction | Flex Day | Tier 1 Instruction | Tier 1 Instruction |
| | | | Or | | |
| C | Tier 1 Instruction | Tier 1 Instruction / Small Group | Tier 1 Instruction / Small Group | Tier 1 Instruction / Small Group | Tier 1 Instruction |
| | | | Or | | |
| D | Tier 1 Instruction | Tier 1 Instruction / Thirty-Minute Intervention | Tier 1 Instruction | Tier 1 Instruction / Thirty-Minute Intervention | Tier 1 Instruction |

*Source: Schuhl et al., 2024, p. 126.*

**Figure 1.3: Tier 1 and Tier 2 schedule options.**

teacher is more concerned about a second-grade student entering third grade who cannot yet add and subtract with an understanding of place value than a student still working to identify quadrilaterals. Both are important, but students are more likely to be successful in mathematics with the former than the latter. A third-grade team continues working with any student still working to identify shapes.

Without identifying essential learning standards as a team, teachers will individually choose their own from year to year. A teacher may teach a lesson that does not go well and independently decide to either move on or reteach it based on what they believe is important for students to learn. As a result, inequities in learning occur. Instead, you and your team collectively choose essential learning standards to guarantee what every student in that grade or course will learn. Use these standards to engage in rich discussions related to instruction, learning, and assessment, and collectively work to ensure the learning of every student. See chapter 3 (page 43) for more information related to essential learning standards for core instruction or the *Mathematics Unit Planning in a PLC at Work* series.

### Rigorous Mathematics

As teachers, we expect every student to learn rigorous mathematics at their grade or course level. *Rigorous mathematics* means teaching grade-level standards with a balance, or equal intensity, of conceptual understanding, application, and procedural fluency (National Governors Association Center for Best Practices and Council of Chief State School Officers [NGA & CCSSO], 2010; Student Achievement Partners, 2015). Students apply their learning to real-world and mathematical tasks as they use different models and strategies to learn concepts and develop efficient and accurate procedures for procedural fluency. Without all three parts of mathematical rigor—conceptual understanding, application, and procedural fluency—students may have gaps in their learning. It is important to address all three components during Tier 1 instruction, with a greater emphasis on conceptual understanding and application in Tier 2.

#### Conceptual Understanding

In *Adding It Up: Helping Children Learn Mathematics*, educators Jeremy Kilpatrick, Jane Swafford, and

Bradford Findell (2001) describe *conceptual understanding* as follows:

> Conceptual understanding refers to an integrated and functional grasp of mathematical ideas. Students with conceptual understanding know more than isolated facts and methods. They understand why a mathematical idea is important and the kinds of contexts in which it is useful. They have organized their knowledge into a coherent whole, which enables them to learn new ideas by connecting those ideas to what they already know. Conceptual understanding also supports retention. Because facts and methods learned with understanding are connected, they are easier to remember and use, and they can be reconstructed when forgotten. (pp. 118–119)

In mathematics, understanding a concept means students make connections from prior learning to current learning, and they know multiple strategies and models to use to solve problems. Students understand the overall concept and are not reliant on choosing a memorized fact or algorithm. Conceptual understanding is critical to both Tier 1 and Tier 2 instruction. Too often, Tier 2 instruction focuses on facts and algorithms that students sometimes replicate in class. However, without conceptual understanding and the ability to answer, *What am I learning?* or *Why am I learning it?*, students often forget the fact or algorithm shortly after the lesson. As a collaborative team, you discuss the concept students are learning in each essential standard and the connections to prior knowledge students need to make. You also identify the different strategies or models you might use to teach the essential standard in a way that builds a conceptual understanding for students.

### Application

Students learn mathematics through application. *Application* means students are solving real-world problems or applying mathematics in various contexts. As students learn mathematics strategies for a concept, they practice and apply that learning in contextual problems. Through real-life contexts, students deepen their understanding of mathematical concepts and make connections within current or previous learning. As students develop procedural fluency, they apply it to real-life mathematics problems to create automaticity with the procedural fluency skill they are learning. Application supports conceptual understanding and procedural fluency.

For example, in middle school, students might be learning how to find a percentage. They learn various strategies and models to find a percentage, such as ratio tables, base ten blocks, and double number lines. They apply that learning to word problems involving real-life percentages like the following: *Pedro bought a new shirt. It originally cost $36 but was on sale for 20 percent off the original price. How much did Pedro pay for the new shirt?*

This real-world application question grows students' understanding of the percentage concept. Over time, students develop procedural fluency and use a percent equation or proportions to solve problems involving percentages. They then use procedural fluency to solve application problems and continue to make sense of the mathematics they are learning.

### Procedural Fluency

*Procedural fluency* means students accurately, efficiently, and flexibly solve tasks that are often demonstrated through mathematics questions related to basic facts and algorithms (NCTM, 2014). It does not mean students have every fact memorized, but they can generate the fact, if needed. For example, an elementary student might just know 8 + 6, or they might use strategies, such as doubles (8 + 8 and subtract 2, or 6 + 6 and add 2, or even see 8 + 6 as 7 + 7), make a ten (8 + 2 = 10 and 10 + 4 = 14, or 6 + 4 = 10 and 10 + 4 = 14), or add on (8, 9, 10, 11, 12, 13, 14). Doubles or make a ten are efficient and flexible strategies students might use when adding numbers and are strongly rooted in number sense.

In high school, students show procedural fluency when they graph $y = 3x + 2$ by placing a point at the $y$-intercept $(0, 2)$, then placing another point using the slope of 3 at $(1, 5)$, and connecting the points to draw a line on a coordinate plane. It is possible some students might also find the $y$- and $x$-intercepts to graph the line by connecting the points $(0, 2)$ and $(-\frac{2}{3}, 0)$.

As your collaborative team plans for rigorous mathematics learning, consider the conceptual understanding, application, and procedural fluency needed for students to demonstrate proficiency. Your team can build a shared understanding of rigorous mathematics for an essential learning standard using a protocol, like the one shown in figures 1.4–1.8 (pages 17–21). These figures show examples of conceptual understanding, application, and procedural fluency for kindergarten, grade 2, grade 4, grade 7, and algebra 2.

For conceptual understanding, identify the overall concept and determine some strategies or models to use to develop the concept. For application, focus on real-world examples to develop mathematics relevance. For procedural fluency, determine the efficient fact or procedure students are expected to do in your grade level or course for the essential learning standard.

Students need to understand what they are learning mathematically and how it fits into their previous and future learning. In other words, what is the big idea, and what strategies and models can be used to develop that understanding? Then, through application and practice, students develop and understand procedural fluency in such a way that they use it to make meaning of future content. Teaching rigorous mathematics in Tier 1 and Tier 2 is critical for accelerating student learning to grade or course level and beyond.

## High-Quality Tier 1 and Tier 2 Instruction

High-quality, effective mathematics instruction is critical to students, whether they receive it in Tier 1 or Tier 2. This book offers numerous strategies you and your team can implement to accelerate student mathematics learning, regardless of the curriculum materials you use. Mattos (2023) often emphasizes, "You cannot intervene your way out of poor instruction." Even when students are not yet at grade level, it is important to incorporate grade-level learning in every lesson to prevent them from needing Tier 2 supports (see chapter 3, page 43, for more information about teaching to grade- and course-level content).

When students require Tier 2 supplemental interventions, implement research-informed instruction and feedback to effectively accelerate student learning.

| **Essential Standard:** K.OA.A.5 Fluently add and subtract within 5. | | |
|---|---|---|
| **Task:** Make different combinations of 5. | | |
| **Conceptual Understanding** | **Application** | **Procedural Fluency** |
| • Adding means putting together or composing, while subtracting means taking apart or decomposing.<br>• Count red and blue crayons (or cubes) to make different combinations of 5. Record combinations using the following blank crayon boxes: | Grandma has a basket with 5 pieces of fruit. Some of the fruits are bananas and some are apples. How many bananas and apples can Grandma have in the basket? Can you make Grandma's fruit basket in more than one way? | • Show 5 using your fingers in two different ways.<br>• I have a tower of 5 cubes, but I hid some behind my back. I am showing you 3 cubes. How many cubes am I hiding? |

*Source for standard: NGA & CCSSO, 2010.*

**Figure 1.4: Example of rigorous mathematics in kindergarten.**

*Visit **go.SolutionTree.com/MathematicsatWork** for a free blank reproducible version of this figure.*

| **Essential Standard:** |||
|---|---|---|
| 2.NBT.A.4 Compare two three-digit numbers based on meanings of the hundreds, tens, and ones digits, using >, =, and < symbols to record the results of comparisons. |||
| **Task:** |||
| Which is greater: 234 or 243? |||
| **Conceptual Understanding** | **Application** | **Procedural Fluency** |
| Numbers can be compared using place value. Compare two quantities beginning with the largest place value.<br>**Base ten blocks** 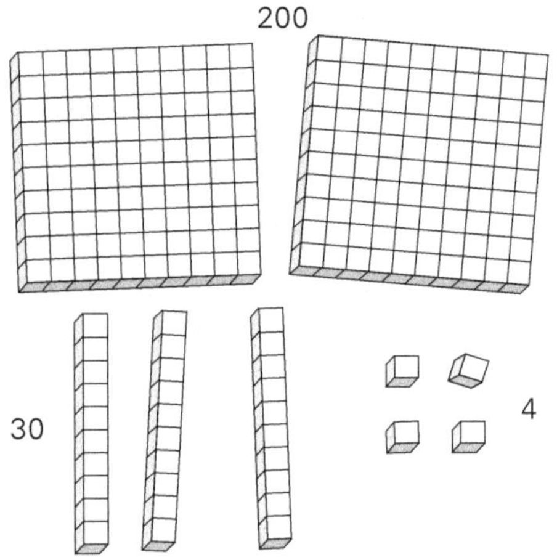 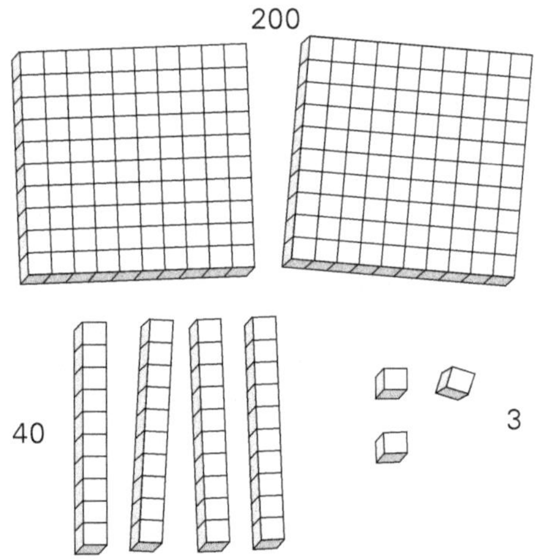<br>200<br>30    4<br>―――――――<br>200<br>40    3<br>Begin by comparing the largest place value, the hundreds place. If the two values in the hundreds place are equivalent, move to the next largest place, the tens. Continue comparing amounts in like places until one amount is greater than or less than the other. | Which brother has more comic books in his collection?<br><br>\| \| Number of comic books \|<br>\|---\|---\|<br>\| Mark \| 234 \|<br>\| Matthew \| 243 \| | Write <, >, or = in the circle to compare the two numbers.<br><br>234  ◯  243 |

*Source for standard: NGA & CCSSO, 2010.*

**Figure 1.5: Example of rigorous mathematics in grade 2.**

| Essential Standard: |||
|---|---|---|
| 4.NBT.B.5 Multiply a whole number of up to four digits by a one-digit whole number, and multiply two two-digit numbers, using strategies based on place value and the properties of operations. Illustrate and explain the calculation by using equations, rectangular arrays, and/or area models. |||
| **Task:** What is 15 × 24? |||
| **Conceptual Understanding** | **Application** | **Procedural Fluency** |
| Multiplication is repeated addition of equal-sized groups.<br><br>15 × 24 can be thought of as 15 groups of 24 objects, which is like a repeated addition problem.<br><br>**Groups**<br><br>(24) (24) (24) (24) (24)<br>(24) (24) (24) (24) (24)<br>(24) (24) (24) (24) (24)<br><br>**Area Model**<br><br>|    | 20  | 4  |<br>|----|-----|----|<br>| 10 | 200 | 40 |<br>| 5  | 100 | 20 |<br><br>200 + 40 + 100 + 20 = **360** | Carmen earns $15 per hour mowing lawns. She worked 24 hours last week. How much did Carmen earn last week mowing lawns? | Partial product algorithm for fourth grade:<br><br>$$\begin{array}{r} 24 \\ \times\ 15 \\ \hline 20 \\ 100 \\ 40 \\ +200 \\ \hline \mathbf{360} \end{array}$$<br><br>Multiplication algorithm is expected in fifth grade:<br><br>$$\begin{array}{r} 24 \\ \times\ 15 \\ \hline 120 \\ +240 \\ \hline \mathbf{360} \end{array}$$ |

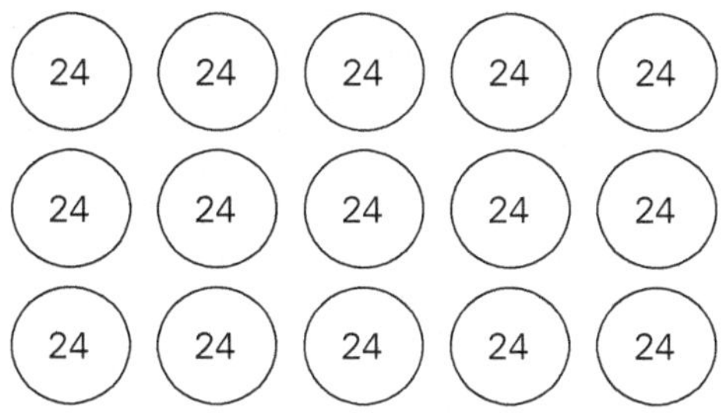

*Source for standard: NGA & CCSSO, 2010.*

**Figure 1.6: Example of rigorous mathematics in grade 4.**

**Essential Standard:**

7.RP.A.3 Use proportional relationships to solve multistep ratio and percent problems. Examples: simple interest, tax, markups and markdowns, gratuities and commissions, fees, percent increase and decrease, percent error.

**Task:**

What is 25 percent more than 145?

| Conceptual Understanding | Application | Procedural Fluency |
|---|---|---|
| Percentages show a ratio out of 100. Proportions can be used to solve percentage problems.<br><br>This means find 25 percent of 145 and add it to 145 or find 125 percent of 145. The answer is greater than 145.<br><br>**Ratio Table**<br><br>| % | # |<br>|---|---|<br>| 100 | 145 |<br>| 10 | 14.5 |<br>| 5 | 7.25 |<br>| 20 | 29 |<br>| 25 | 36.25 |<br>| 125 | (181.25) |<br><br>**Proportion** $\frac{125}{100} = \frac{x}{145}$<br><br>**Double Number Line**<br><br>0%  25%  50%  75%  100%  125%<br>0   36.25  72.50  108.75  145  (181.25)<br><br>**Tape Diagram**<br><br>| 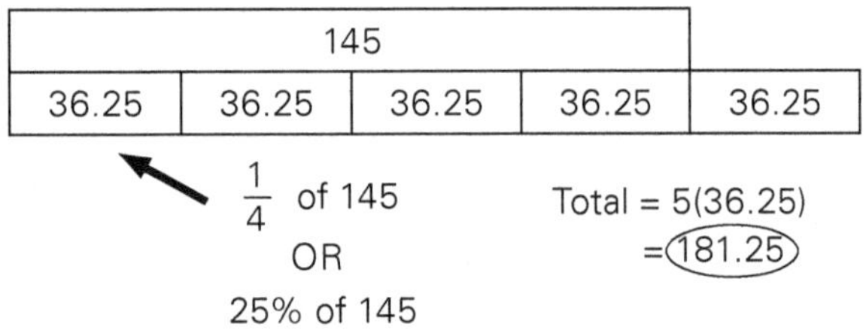 145 | |<br>|---|---|<br>| 36.25 \| 36.25 \| 36.25 \| 36.25 \| 36.25 | |<br><br>$\frac{1}{4}$ of 145  Total = 5(36.25)<br>OR              = (181.25)<br>25% of 145 | Jamie is selling a painting. She initially priced the painting at $145 and is now selling it for 25% more than the initial price. What is the current price of the painting? | **Percent Equation**<br><br>1.25(145) = (181.25)<br><br>OR<br><br>0.25(145) = 36.25<br>36.25 + 145 = (181.25)<br><br>**Proportion** $\frac{125}{100} = \frac{x}{145}$ |

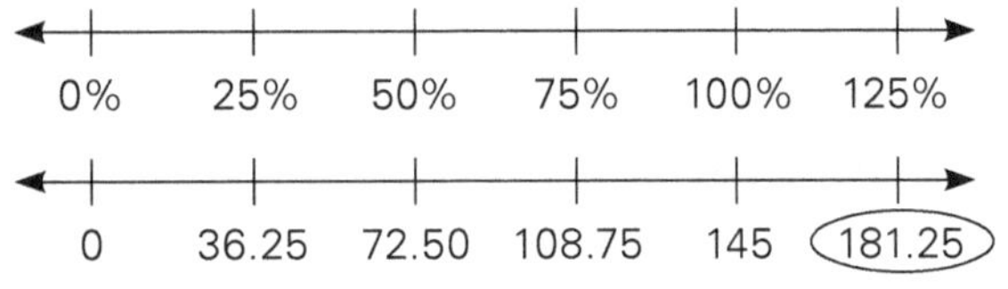

*Source for standard: NGA & CCSSO, 2010.*

**Figure 1.7: Example of rigorous mathematics in grade 7.**

| **Essential Standard:** |||
|---|---|---|
| HSF-BF.B.3 Identify the effect on the graph of replacing $f(x)$ by $f(x) + k$, $kf(x)$, $f(kx)$, and $f(x + k)$ for specific values of $k$ (both positive and negative); find the value of $k$ given the graphs. Experiment with cases and illustrate an explanation of the effects on the graph using technology. |||
| **Task:** |||
| How do the graphs $f(x) = \sin x$ and $g(x) = 2\sin(x - \pi) + 4$ compare? |||
| **Conceptual Understanding** | **Application** | **Procedural Fluency** |
| Functions can be moved using transformations:<br>• Translation (up, down, right, left)<br>• Dilation (stretched, compressed)<br>• Reflection<br>• Rotation<br><br>Sine and cosine functions model periodic phenomenon.<br><br>**Table of Values**<br><br>| $x$ | $f(x)$ | $g(x)$ |<br>|---|---|---|<br>| 0 | 0 | 4 |<br>| $\frac{\pi}{2}$ | 1 | 2 |<br>| $\pi$ | 0 | 4 |<br>| $\frac{3\pi}{2}$ | –1 | 6 |<br>| $2\pi$ | 0 | 4 |<br><br>**Graphs**<br><br>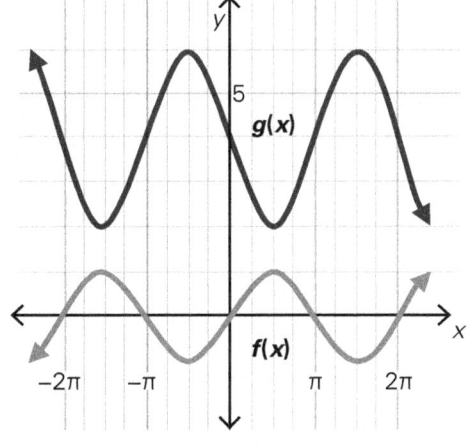 | Clara built a small wind turbine. The blades of the turbine are 2 feet long, and the center of the blades is 4 feet off the ground. She notices it takes $2\pi$ seconds for one full rotation of the blades as measured from a horizontal blade moving clockwise. What is a model for the function created by the rotating blades? How does the model differ from $y = \sin x$? | $f(x)$ and $g(x)$ are both sinusoidal and have a period of $2\pi$ since in both functions the term $x$ has a coefficient of 1. $f(x)$ is the sine parent function with a midline at $y = 0$ and amplitude of 1. $g(x)$ is shifted right $\pi$ and up 4 from $f(x)$. The midline of $g(x)$ is at $y = 4$ and the amplitude is 2, so $g(x)$ is stretched by 2. This all means $g(x)$ has the same period as $f(x)$, twice the amplitude, and is shifted right $\pi$ and up 4. |

*Source for standard: NGA & CCSSO, 2010.*

**Figure 1.8: Example of rigorous mathematics in algebra 2.**

Remember to determine the effectiveness of any Tier 2 intervention as a collaborative team. If an instructional strategy or model was used that grew student learning, consider adding it to the unit plan. By doing so, you can incorporate the new strategy into Tier 1 core instruction the next year as a prevention method.

Figure 1.9 shows a completed example of a template your team can use for planning Tier 2 interventions. The template focuses your team on a specific essential standard and the progression of learning required for students to successfully master that standard. Notice the planning tool starts by recognizing the procedural fluency students ultimately need and then has your team identify the conceptual understanding and application needed to develop the desired procedural fluency.

We built our Mathematics in a PLC at Work framework for collaborative teams on the premise that every student can learn mathematics. Working together with strong Tier 1 and Tier 2 strategies and models for instruction in a collaborative team makes it possible for every student to learn mathematics.

## Conclusion

When students have not yet learned grade- or course-level mathematics, the response is not to blame students, blame previous teachers, or solve the issue by ability grouping students for initial core instruction. Instead, your team works together to address the PLC at Work four critical questions for the most essential standards. You and your collaborative team ensure students learn grade- or course-level standards through intentional instruction and assessment in Tier 1 and Tier 2, making sure to designate time during the school day for Tier 2 learning each week. Together, you and your team accelerate student learning to grade or course level in a rigorous mathematics program.

## Questions to Consider for Next Steps

As a collaborative team, use the following questions to reflect on your current practices and determine any next steps related to working as a collaborative mathematics team in a PLC at Work focused on student learning.

1. How do you and your team plan units of instruction that focus on the essential learning standards? (See the *Mathematics Unit Planning in a PLC at Work* series for ideas to support planning units of instruction..

2. How do you currently address interventions and extensions during daily lessons when you see students aren't learning or struggling to learn the target?

3. How does your team address interventions and extensions after a common mid-unit assessment? Does this occur during Tier 1 core instruction or Tier 2 intervention?

4. How does your team currently create targeted interventions and extensions during designated Tier 2 time in the school day?

5. How might you and your team create Tier 2 time in the day to re-engage students in learning mathematics if the time is not already included in your master schedule?

6. Which parts of rigor (conceptual understanding, application, and procedural fluency) are you currently teaching in Tier 1 and Tier 2 instruction?

7. How do you and your team keep records of effective instructional practices used in Tier 1 or Tier 2 so they can be used later in the year or the following school year?

**Directions:** Work as a collaborative team to identify the grade- or course-level essential standard and prior knowledge vertical standards that support and lead to learning the essential standard. Next, determine the mathematical rigor of the essential standard by clarifying the required procedural fluency, as well as the conceptual understanding students must develop and examples of its applications. Identify example tasks for each. Finally, determine differentiation and routines to use when teaching the intervention plan. (Note: Many tasks and strategies could also be used during Tier 1 core instruction.)

| Concept or Unit: Multidigit Addition and Subtraction Intervention |||
| --- | --- | --- |
| **Grade or Course Essential Standard:** *State the current essential standard or learning target.* |||
| 4.NBT.B.4 Fluently add and subtract multidigit whole numbers using the standard algorithm. |||
| **Prior Knowledge Standards:** *List the prior knowledge standards that are vertically connected to the essential standard.* |||
| 3.NBT.A.2 Fluently add and subtract within 1,000 using strategies and algorithms based on place value, properties of operations, and/or the relationship between addition and subtraction. <br> 2.OA.B.2 Fluently add and subtract within 20. |||
| **Mathematical Rigor** |||
| **Procedural Fluency Connections** | **Conceptual Understanding Connections** | **Application and Problem Solving** |
| **Purpose and Goals** <br> Grade 4 students need to have procedural fluency with the addition and subtraction standard algorithm. Understanding algorithms based on place value and knowing sums and differences within 20 are needed for students to add and subtract with the standard algorithm. | **Purpose and Goals** <br> Students have worked with addition and subtraction concepts since kindergarten. There are many models and strategies to use to add and subtract. Use base ten blocks, expanded forms of numbers, and place value to connect the standard algorithms to previously learned models and strategies in recent grades. | **Purpose and Goals** <br> Addition and subtraction are operations used in real-world experiences. Students practice the standard algorithms to solve word problems (one or more steps). |
| **Tasks** <br><br> **Entry task:** Engage students in the *Addition and Subtraction Number Talks Within 20*\* task. (For Number Talks strategy, see page 78.) Use the different activities to practice addition and subtraction fact fluency. <br><br> **Follow-up task:** Use the *Tic-Tac-Toe Addition and Subtraction* tasks (Mental Strategies, Over and Under Estimating, and Standard Algorithm)\* to practice solving problems with varied number ranges. Students work in pairs and choose three problems to answer that are in a row. When they complete the row, they check their answers with the teacher. The teacher lets them know if they win or need to continue working. <br><br> If a team wins, have students continue the *game* by asking the team to find another three in a row until the Tic-Tac-Toe has been completed. (For Centers and Games strategy, see page 66.) | **Tasks** <br><br> **Entry task:** Use the *Picture This*\* task. Students work in pairs to make sense of the picture and then write the equation and answer using the expanded and standard algorithms. Some students may need to use base ten blocks to make sense of the task. <br><br> **Follow-up task:** Use the *Three Ways*\* task. There are three different word problems and a fourth task with part of the expanded algorithm shown (students finish the problem and create a word problem as one of their three ways). <br><br> Each task is in a *graphic organizer*, and students create an equation to match the word problem (first three tasks) and then work in groups or move to stations in the room with a partner to complete each part of the graphic organizer. (For Mathematics Graphic Organizers strategy, see page 124.) <br><br> Students are learning the different strategies for adding and subtracting and seeing how they connect to the standard algorithm. | **Tasks** <br><br> **Entry task:** Use the *How Much?*\* task, which utilizes the *Provide the Answers* strategy (see page 146). Answers to each word problem are given and students solve an addition or subtraction equation using the standard algorithm. <br><br> **Follow-up task:** Use the *Leveled Tasks: Addition and Subtraction*\* task. (For Leveled Tasks strategy, see page 49). Differentiate and have students work in pairs on the tasks that they can access and work to grade level or extension, or give pairs the developing task and have them work their way to extension. <br><br> Ask students to solve the tasks using the standard algorithm for addition or subtraction. |

**Figure 1.9:** Essential standard intervention planning tool: Fourth-grade multidigit addition and subtraction example.

continued ▶

| Routines and Differentiation | | | |
| --- | --- | --- | --- |
| **Focus** *What is the focus for the intervention?* | **CRA** *What concrete objects, representational pictures, or abstract equations can be used in the intervention?* | **Differentiation** *What differentiation may be needed when students are not yet learning?* | **Tier 3 Supports** *What Tier 3 supports might students also need?* |
| • Teachers can choose a combination of the tasks in this plan throughout the duration of the intervention. The focus of each task is students understanding adding and subtracting using place value and the standard algorithm.<br>• Students work in pairs or groups of four while completing tasks and use *sentence frames* to have meaningful discourse with one another. (For Sentence Frames strategy, see page 133.) | **C**–Use tools or manipulatives, such as base ten blocks or place value charts with counters (possibly ten frames), to add and subtract with and without regrouping.<br>**R**–Use drawings of base ten blocks and open number lines to show addition and subtraction. Reference the *Visual Progression of Strategies for Addition and Subtraction*\* tasks. (See figures 7.1, page 100, and 7.3, page 102.)<br>**A**–Add and subtract using an expanded algorithm (by place value) and the standard algorithm. | • If students are not yet adding and subtracting with multidigit numbers, start with a two-digit and one-digit number first.<br>• Some students might need to start with multidigit addition and subtraction problems that do not require regrouping.<br>• Use graph paper or an algorithm graphic organizer so students line up each number's digits by place value.<br>• Add and subtract with base ten blocks and write the addition and subtraction in the format of the standard algorithm. | • Practicing fluency strategies with sums and differences within 10 and 20 will move students away from counting by ones and toward understanding the place value system.<br>• Composing and decomposing numbers in different ways will support thinking while transitioning to more efficient strategies.<br>• Modeling the *Known Ten* strategy will aid students in solving addition and subtraction facts within 20 (see figure 7.2, page 101). Use the *Known Ten Interactive Number Talks*\*. |

\**Supporting activities are included in the online reproducibles.*

*Source for standards: NGA & CCSS, 2010.*

*Visit* **go.SolutionTree.com/MathematicsatWork** *to download a free blank reproducible version of this figure, along with completed examples of seventh grade and high school (algebra 1) intervention plans with supporting activities.*

It was November, and Mona was coaching a third-grade team on lesson design. During one of her previous visits, the team focused on lesson design elements to include prior knowledge routines, choice of high- and low-level-cognitive-demand mathematical tasks, and student-led closure routines. On the first day of Mona's coaching visit, two third-grade team members co-created a lesson and discussed how to implement it. They selected a number talk for the opening task, several mathematical tasks balanced in cognitive demand for the heart of the lesson, and a closure prompt to give students at the end of the lesson.

On the second day of the coaching visit, Mona observed students in both team members' classrooms as each teacher taught the lesson. In the first classroom, she noted that students were sitting in rows, and in the second classroom, she noted that students were sitting in groups of three. Mona quickly observed a difference in student interaction and learning between the two classrooms. In the first classroom, only a few students talked to each other when working on tasks. Most students completed the tasks individually and did not provide or receive feedback from their peers. Additionally, some students waited for the teacher to come around and help them before starting a task.

However, in the second classroom, students were expected to talk to each other, use sentence frames to start mathematical conversations, and listen to each other to understand new ways of thinking. Students knew they needed to interact with each other and not wait for the teacher to get them started. During the closure routines, students in the second classroom felt confident about the learning target and there was clear evidence they had met the lesson's target.

# CHAPTER 2

# Build a Community of Learners

*People will forget what you said, people will forget what you did,
but people will never forget how you made them feel.*

—Maya Angelou

Even though both teachers on the team were teaching the same lesson during Mona's observations (see story on page 26), there was a significant difference in student engagement. The second classroom, with its established routines, fostered a community of learners where students felt comfortable learning from each other and sharing their thinking. This environment led to increased engagement, and the students were able to make sense of the mathematics. In contrast, the first classroom placed more emphasis on the teacher as the primary source of learning, resulting in a decreased sense of the classroom being a community of learners. It is important to not only consider the mathematical tasks and instructional routines you choose as a team, but also the strategies used to create a community of learners.

PreK–12 teachers make many decisions that contribute to or hinder building a classroom community focused on learning. Decisions are often related to physical aspects of the room, such as furniture placement, learning displayed on the walls, strategically placed and visible learning targets, and a clutter-free and clean space. Teachers also make decisions related to identifying and teaching expectations for how students will interact with one another and with the teacher during whole-group and small-group instructional experiences. Many decisions are made during Tier 1 and Tier 2 instruction that minimize transition time and maximize learning time with feedback. Building a community of learners is not always easy, but it is always intentional.

## The *Why* and the *What* of Building a Community of Learners

Imagine a classroom community where all students see themselves as mathematicians. What do you see? What do you hear? To build such a community of learners, consider how to develop students' mathematical agency, identity, and positive self-efficacy, as described in table 2.1 (page 28).

Classroom communities centered on building students' positive mathematics identity, agency, and learning are places where students feel the following:

- My thoughts and opinions are accepted.
- I can take risks without fear of being ridiculed
- I can make mistakes because they are seen as opportunities for growth.
- I can be myself and ask questions.
- I can grow my learning of mathematics. (Venturis, n.d.)

Professor of educational leadership Karen F. Osterman (2023) shares that students who have more positive attitudes toward themselves and others interact with the teacher and students in positive and supportive ways. Students tend to generate positive attitudes toward themselves in mathematics when they understand how they learn and feel competent as mathematical learners—they develop positive self-efficacy. Positive self-efficacy is both grown in and is a contributor to a strong classroom community.

Table 2.1: Definition of Mathematics Identity, Agency, and Positive Self-Efficacy

| Mathematics Identity | Students' mathematical identity is how they see themselves in the mathematics classroom—their relationship with the subject of mathematics and mathematical activities.<br>*Students ask: Am I a mathematician? Do I believe I can do mathematics?* |
|---|---|
| Mathematics Agency | Mathematics agency is students' capacity to see themselves as mathematical thinkers who develop rich mathematical understanding and become doers of mathematics.<br>*Students ask: What have I learned that makes sense to apply to this task?* |
| Positive Self-Efficacy | Positive self-efficacy is students' belief in their ability to succeed in achieving an outcome. When students feel competent in mathematics, they are motivated to persevere.<br>*Students ask: I can do this task; what else should I try?* |

*Source: Adapted from NCSM, 2022.*

Developing a powerful classroom learning community requires daily work and unwavering commitment. Students feel they are in a place where they are welcome and belong. *Belonging* is an essential human need, defined as feelings of connection and acceptance, or fitting in (Darling-Hammond et al., 2020; Maslow, 1943). Research shows that a student's sense of belonging is positively correlated with student outcomes such as motivation, interest in the content, and the intent to pursue college or persist in a discipline (Darling-Hammond et al., 2020; Osterman, 2023). Belonging is particularly important for historically marginalized students, who are more likely to experience exclusion and uncertainty about whether they belong in certain contexts, including at school (Darling-Hammond et al., 2020). A sense of belonging is an essential precursor to learning, especially in mathematics.

So, how can your team foster a classroom culture focused on making connections and creating a sense of belonging? A University of Cambridge (2016) study finds that building relationships with adolescents increases student learning and altruism and decreases problem behaviors interfering with learning. NCSM (2022) states that one of the most powerful actions teachers take in their mathematics classrooms is getting to know students—in other words, building relationships through mathematics.

Use the chart in figure 2.1 to evaluate how connected students feel in your classroom and how well they feel

**Directions:** Read the characteristics of belonging statements. Then reflect on your current practices and next steps.

| Characteristics of Belonging | Reflection<br>Current Strengths? Possible Next Steps? |
|---|---|
| Students' thinking and approaches to mathematics are respected, valued, and accepted by their teachers and peers. | |
| Students see themselves in the mathematics they are learning and see its relevance in their lives, which grows a positive mathematics identity. | |
| Students have the confidence to share what they think about a topic or a problem, rather than feel like they need to respond in a way the teacher or textbook expects. | |
| Students can succeed even if they make mistakes or run into obstacles (in other words, they have positive self-efficacy and a growth mindset), and their teachers have high expectations for them to grow while providing the support to experience success. | |
| Students have agency in mathematics—they feel empowered to contribute to and shape their mathematics learning environment. | |
| Students own their mathematical ideas because they have reasoned them through rather than memorized information. | |

**Figure 2.1: Reflection tool on characteristics of belonging.**

*Visit* **go.SolutionTree.com/MathematicsatWork** *for a free reproducible version of this figure.*

they belong. This is a reflection tool on the characteristics of belonging (Moeller, n.d.).

To help each student develop a sense of belonging, agency, identity, and positive self-efficacy, you and your team intentionally create mathematics learning environments during both Tier 1 and Tier 2 instructional time that actively include and value each student's contribution, reasoning, and learning strategies. Doing so brings joy to learning and teaching mathematics. To accomplish such a learning environment as a collaborative team, consider the actions featured in table 2.2.

To create a community of learners, you and your team collaborate to define the specific characteristics you are developing in your mathematics students. Once you identify these characteristics, work together to establish deliberate structures, protocols, and strategies that provide students with opportunities to grow their mathematics identity and build confidence, agency, and self-efficacy. By agreeing on the characteristics and protocols to be used, students will have the chance to continue developing these qualities when receiving targeted intervention in a colleague's classroom.

Figure 2.2 shows an example of characteristics you and your team may want students to develop when learning mathematics. Consider what your collaborative team includes in a portrait of a mathematician.

Table 2.2: Build a Community of Learners in Tier 1 and Tier 2

| Build a Community of Learners in Tier 1 and Tier 2 ||
|---|---|
| <ul><li>Teach, model, and reinforce routines for students to embrace learning from errors as well as listen to and learn from their peers.</li><li>Reinforce routines that promote a positive mathematical identity, mathematics agency, and self-efficacy.</li><li>Create intentional, positive social interactions to ensure a strong sense of belonging.</li></ul> ||
| **Build a Community of Learners in Tier 1** | **Build a Community of Learners in Tier 2** |
| <ul><li>Create intentional opportunities to get to know students, their strengths, and their interests.</li><li>Create a classroom culture where students are expected to learn from each other.</li><li>Use asset-based language and team actions to grow every student.</li></ul> | <ul><li>Focus on identifying student strengths and building on them.</li><li>Ensure that the Tier 2 instruction environment is a safe place for students to take risks and learn from each other.</li><li>Create Tier 2 interventions as opportunities for students to learn rather than punishment for not having learned yet.</li></ul> |

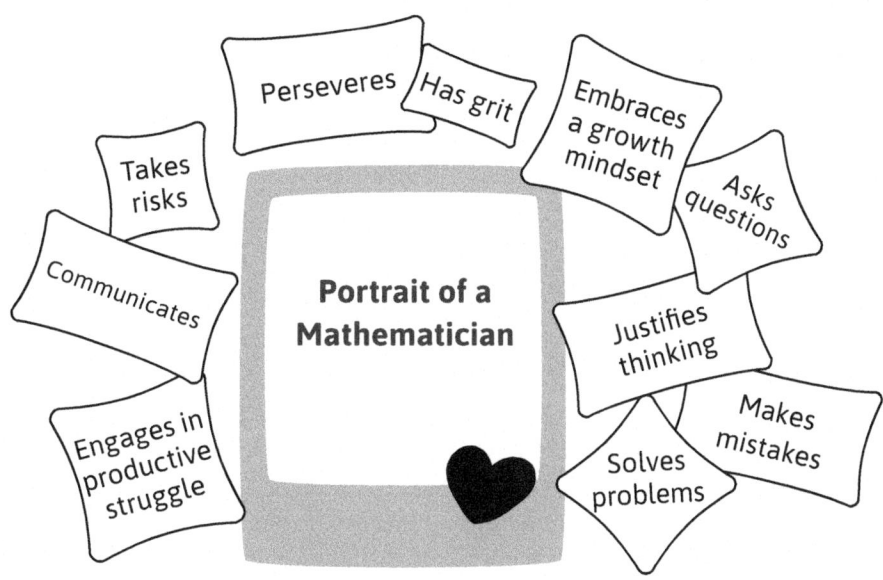

*Source: © 2022 by Mona Toncheff. Used with permission.*

**Figure 2.2:** Portrait of a mathematician.

Discuss what the actions look and sound like once your team agrees on what you want your community of learners to do when learning mathematics. The following strategies are designed to support building such a community in your Tier 1 and Tier 2 learning experiences.

## The *How* of Building a Community of Learners

Undeniably, it is crucial to be intentional about structuring social interactions during Tier 1 and Tier 2 instruction. Developing desired social interactions contributes to establishing a culture of learning. Consider how to build strong mathematical mindsets and create an environment where students see mistakes as learning opportunities. To achieve productive social interactions, it is important to clarify expectations and intentionally use protocols and structures during small-group and whole-group discourse in the classroom with students (see chapter 9, page 129, for more information on classroom discourse). Establishing a community of learners starts with establishing clear norms and expectations in the classroom.

### Norms for Student Interactions

When Mona was a new high school mathematics teacher, she assumed students knew how to behave as high school mathematicians; however, she quickly learned they did not. She had to clarify and teach students classroom norms and behavior expectations for whole-group, small-group, or independent learning. More important than describing behaviors, the classroom norms helped students see themselves as mathematicians.

Other names for classroom norms include rights and responsibilities, classroom commitments, or classroom expectations. Regardless of the name, you and your students need structures to support meaningful mathematical discourse, engagement, and action.

To help students see themselves as mathematicians, start by creating clear and concise student expectations for both Tier 1 and Tier 2 instruction. These norms clarify the learning behaviors that are important in your own classroom, your colleagues' classrooms, in future mathematics classrooms, and even apply to life beyond preK–12 education. What you expect from students and what students expect from each other during learning experiences emphasizes lifelong learning and models how students can continue to grow socially and in their mathematics interactions with one another (Hierck, 2017).

Some norms clarify what it means for an individual student to be a mathematician. What are the behaviors expected when learning mathematics? Figure 2.2 (page 29), for example, shared a portrait of a mathematician. For each behavior or action in the portrait, discuss with your team what you need to see or hear that demonstrates students are developing the behavior. For example, what do you need to see and hear if a student is persevering or solving problems? Next, discuss how you will teach the expectations for each behavior. Additionally, you might ask students to share their insight into what it takes to be a mathematician.

Once you have clarified the behaviors needed for students to be mathematicians, also clarify how students will interact and learn from one another in a collaborative classroom. Armed with your list of what it looks and sounds like for students to become mathematicians and work collaboratively in the classroom, identify the norms needed for students to behave accordingly during Tier 1 and Tier 2 instruction.

Create posters or anchor charts of classroom norms to share with students or co-create the norms with students. Figure 2.3 shows an example of norms written as rights and responsibilities for students in a learning community, and figure 2.4 shows an example of student behavior norms and skills for students working in groups during Tier 1 and Tier 2 learning experiences.

| Rights | Responsibilities |
|---|---|
| Each student has the right to: | Every student has the responsibility to: |
| Ask for help | Help others |
| Be heard | Listen to others |
| Make a mistake | Learn from mistakes by acting on feedback |
| Express thoughts about solution pathways | Be open to embracing new ways of thinking |
| Disagree with respect | Seek consensus within the team |
| Learn | Learn with everyone on the team |

*Source: Adapted from Toncheff et al., 2024.*

**Figure 2.3: Sample classroom discourse rights and responsibilities.**

| Small-Group Norms | What This Looks Like and Sounds Like |
|---|---|
| Students should:<br>• Listen carefully and respect one another<br>• Contribute to the assigned team task<br>• Ask other team members for help when needed<br>• Help other team members who ask<br>• Insist on logical explanations before changing the pathway of the team solution | Students will:<br>• Use quiet, conversation-level voices<br>• Stay engaged and persevere through the mathematical task assigned<br>• Ask peers for help, and then ask the teacher<br>• Support each other<br>• Ask for reasons or ask each other questions<br>• Criticize ideas but not other students<br>• Have a sense of humor |

*Source: Toncheff et al., 2024, p. 75.*

**Figure 2.4: Sample classroom norms for working in groups.**

Figure 2.5 shows another example of norms for a middle school classroom. The team clarified their classroom expectations with students, and then the teachers caught students modeling these behaviors, took photos, and showed the desired behaviors on the poster using the pictures. Each teacher on the team then created posters in their respective classrooms so students could see themselves and their peers positively modeling the expectations.

In the primary grades, a poster of norms written with words is not very helpful. Instead, consider using pictures with very brief phrases, like the poster in figure 2.6 (page 32). This shows students the expectations, ideally with pictures of students themselves.

In preK and kindergarten, you might consider taking pictures of students modeling desired behaviors and then post the pictures in the classroom to remind students of each expectation, as applicable, throughout the day. For example, if it is time for students to get mathematics bags from their cubbies and bring them to the carpet, show a picture of a student from the classroom sitting with their legs crossed on the carpet, hands folded in their laps, and the bag sitting neatly in front of them. If students are expected to work with a partner, show two students working together, leaning forward to depict using their inside voices. Periodically change pictures so every student sees themselves as exemplary during the year.

After establishing norms or community expectations to support a safe and respectful classroom setting for students to learn from one another, consider how your physical learning environment communicates what you value in a mathematics classroom. How the classroom is organized and structured will aid students in discussing, sharing, and learning from their peers.

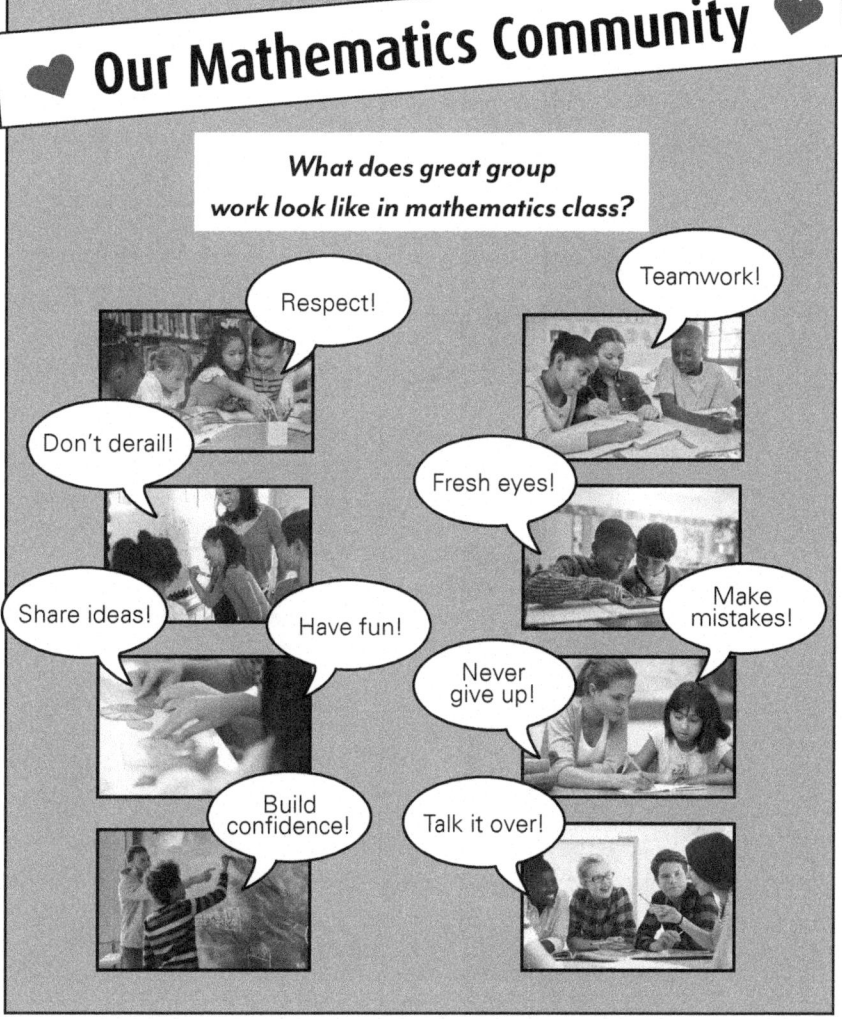

*Source: Adapted from © 2023 by Seventh-Grade Team, Emily Gray Junior High, Tanque Verde, Arizona.*

**Figure 2.5: Example of classroom community expectations.**

# CLASSROOM ROCKET EXPECTATIONS

**R is for Respect**
- Think and reflect on how your words and actions impact others.
- Use materials appropriately.

**O is for On-Task**
- Have materials ready.
- Use time wisely.
- Stay focused and follow directions.

**C is for Calm Body**
- Keep your hands and feet to yourself.
- Stay in your own personal space and spot.
- Use an appropriate volume and tone of voice.

**K is for Kindness**
- Think before you act or speak.
- Use encouraging words.

**E is for Engaged**
- Participate in class.
- Listen actively.
- Show your best effort.

**T is for Teamwork**
- Display a positive attitude.
- Help and share with others.
- Include everyone.

*Source: Adapted from © 2023 by Ashley Mistretta. Used with permission.*

**Figure 2.6: Classroom rocket expectations.**

### Learning Community Environment

The classroom layout plays an important part in building a learning community. An open, collaborative classroom culture, one that encourages respectful and positive interactions, supports preK–12 students in thinking about their own and classmates' learning (National Council for Curriculum and Assessment, 2022). Ideally, the physical environment allows teachers and students to move easily throughout the room and includes space for students to share their thinking with peers.

The physical environment in a mathematics classroom includes the objects teachers and students can touch and use during mathematics lessons as well as the furniture layout (in essence, the intentional space created around the room). The physical environment includes, but is not limited to, student desks and materials, manipulatives, whiteboards, anchor charts, and any presentation system to share student thinking. You and your colleagues work together to purposefully organize the classroom's physical space.

How your classroom space is organized communicates to students what you value. For example, desks organized into groups show a value in and a belief that students should interact and learn from one another. Desks in rows show a belief that the teacher is the authority in the classroom and students learn from independent work. Figure 2.7 (page 34) shows examples of student desk arrangements and questions for your team to discuss related to the messages these arrangements send to students.

Student motivation research shows classroom environments that are both physically and emotionally safe facilitate social, emotional, and academic growth (Hamre & Pianta, 2005; Zins & Elias, 2007). A positive, supportive learning environment with clear social and emotional expectations that promote social interactions provides opportunities for students to own their learning (Berlin & Cohen, 2020). Figure 2.8 (page 35) lists some ideas to consider when organizing physical space and objects in the classroom.

As your team clarifies expectations for students to learn from one another and engage in meaningful discourse, consider how to use flexible seating structures. Determine whether to put students in groups of two, three, or four while they work through mathematical tasks. In the primary grades, we recommend groups of two (pairs).

The physical environment students encounter in the classroom should be inviting and encourage them to develop a strong mathematics identity in which they see themselves as capable mathematicians. Consider how to make the mathematics classroom a physically and emotionally safe environment where students take risks to learn more. In both Tier 1 and Tier 2 learning experiences, encourage students to embrace their mistakes and learn from them to deepen their understanding and application of mathematics.

### Mistakes as Learning Tools

Whether Sarah was teaching algebra 1 or working in the kindergarten classroom, she noticed that students quickly erased all their work as soon as they knew their solution pathway was incorrect instead of examining which part might be on the right track and solving again from that point. Past NCTM president Linda M. Gojak (2013) says, "Helping students to learn from their mathematical mistakes can give us insight into their misconceptions and, depending on our instructional reactions, can enable them to develop deeper understanding of the mathematics they are learning."

If you want students to learn from their mistakes and understand the power of *not yet*, it is important to teach, model, and reinforce routines that embrace mistakes. Students who learn from their mistakes understand that those mistakes show what they know and what they still need to learn.

Tracy Zager (2017) shares three specific steps for using mistakes as a tool for students to examine and refine their mathematical thinking:

1. Teach students to make mistakes in stride.
2. Teach students to keep going when they realize they made a mistake! They need to get to the bottom of their misunderstanding with tenacity, determination, and curiosity until they understand exactly why their reasoning is flawed.
3. Teach students to make the most of their knowledge and experience they gained by figuring out their mistake. Now they can develop better reasoning and improved methods that yield results. (pp. 57–58)

**Directions:** Look at the four classroom desk arrangements, and then answer the reflection questions. Discuss your answers as a collaborative team.

| Picture 1 | Picture 2 |
|---|---|
|  |  |
| Picture 3 | Picture 4 |
|  |  |

**Reflection Questions**

1. Which classroom setup or setups communicate that student-to-student discourse is a requirement? Explain why.

2. Which classroom or classrooms have ample space for the teacher and students to move around the room? Explain why.

3. Which classroom or classrooms have space for small-group instruction?

4. How could the space in each classroom be used for Tier 2 instruction?

5. Which classroom or classrooms have ample space for students to display their thinking around the room during the lesson?

*Source: © 2023 by Mona Toncheff. Used with permission.*

**Figure 2.7: Desk arrangement reflection.**

*Visit **go.SolutionTree.com/MathematicsatWork** for a free reproducible version of this figure.*

**Directions:** Reflect on how your current practices compare to the ideas in this chart. Then determine your next steps to strengthen learning through your use of physical spaces and objects throughout Tier 1 and Tier 2 instruction.

| Element of Physical Environment | Questions to Consider | Reflection and Next Steps |
|---|---|---|
| Whiteboards and wall space dedicated to student work | • How are whiteboards in the classroom used to share student thinking?<br>• How is wall space used to display artifacts of student learning? | |
| Student seating and other learning spaces (for example, carpet time in elementary grades) | • What opportunities for mathematical engagement exist in each area of the classroom?<br>• How does the desk or table arrangement encourage students to learn from one another? | |
| Numeracy-rich environment | • How are anchor charts displayed in the classroom? How are mathematics word walls used to develop mathematical language?<br>• In the primary grades, how do you help students access mathematics in literature to allow them to extend and expand their mathematical thinking? | |
| Space for whole-group, small-group, and independent work | • How do you ensure there is room for students to move comfortably throughout the classroom?<br>• How are students able to interact and learn from one another in whole-group and small-group learning?<br>• How well can students work on their own, when needed? | |
| Mathematics manipulatives | • Where can you place the mathematics tools students need so they are easily accessible?<br>• What is the best location for students to access materials (for example, markers, colored pencils, manipulatives, handheld technology)? | |

*Source: Adapted from Michigan State University College of Natural Science, n.d.*

**Figure 2.8: Physical environment considerations in a learning community.**

*Visit **go.SolutionTree.com/MathematicsatWork** for a free reproducible version of this figure.*

Research finds that students who analyze both correct and incorrect student work improve their understanding when the worked-out problems are detailed, align to a clear learning target, and highlight steps needed to solve the problem (Barbieri, Miller-Cotto, Clerjuste, & Chawla, 2023). To model the routine of analyzing errors and learning from them, use error analysis tasks. Create tasks designed to show students' common misconceptions in their work to help them learn from their mistakes. Figures 2.9–2.13 (pages 36–38) show sample error analysis tasks.

You might also find high-quality, multiple-choice items aligned with the essential standard you are teaching. You might then give students the multiple-choice task during a warm-up or class instruction and ask them to determine, for example, why B is the incorrect answer. Students will often learn more about mathematics from determining why a distractor is incorrect than in finding the correct answer.

Mistakes are a mirror into student thinking. During both Tier 1 and Tier 2 instruction, teachers focus more on student thinking and less on whether a problem is wrong or right. When there are trends in student errors, consider creating focused Tier 2 intervention groups in which you and your colleagues shuffle and deploy students to address the targeted thinking that created the error. Use more prompts and questions to help students develop better reasoning and a deeper understanding of mathematics. Teachers and students can use prompts and questions like the following.

- "Tell me more."
- "What makes you say that?"
- "_____ just said something interesting. Let's explore it a little more."
- "Hmmm. You bring up something interesting. What do others think about that?"

More examples of questioning techniques can be found in chapter 9 (page 129). We reference embracing mistakes and error analysis throughout this book because it is a powerful way to engage students in thinking and learning through discourse. Additionally, learning through error analysis helps students understand that learning happens when mistakes are made and then understood. Building a community of learners requires students to see mistakes as learning opportunities.

---

**Error Analysis Task: Kindergarten**

**Directions:** Read the task and the student work. With your partner, identify the error in the student work and describe why it is an error. Rework the problem correctly.

**Task:** There are 6 candles on my birthday cake. I blew on the cake and 4 candles went out. Some of the candles stayed lit. How many candles do I still need to blow out?

**Student response:**

| Describe the Error | Rework the Problem |
|---|---|
|  |  |

---

**Figure 2.9: Kindergarten error analysis task.**

*Visit **go.SolutionTree.com/MathematicsatWork** for a free blank reproducible version of this figure.*

**Error Analysis Task: Grade 1**

**Directions:** Read the task and the student work. With your partner, identify the error in the student work and describe why it is an error. Rework the problem correctly.

**Task:** Write the time on the digital clock to match the time on the analog clock.

**Student response:**

[Analog clock showing hour hand between 3 and 4, minute hand on 6]

3:30

| Describe the Error | Rework the Problem |
|---|---|
| | |

Figure 2.10: Grade 1 error analysis task.

---

**Error Analysis Task: Grade 3**

**Directions:** Read the task and the student work. With your partner, identify the error in the student work and describe why it is an error. Rework the problem correctly.

**Task:** Jorge walks $\frac{3}{4}$ of the way home. Represent the situation on the bar diagram.

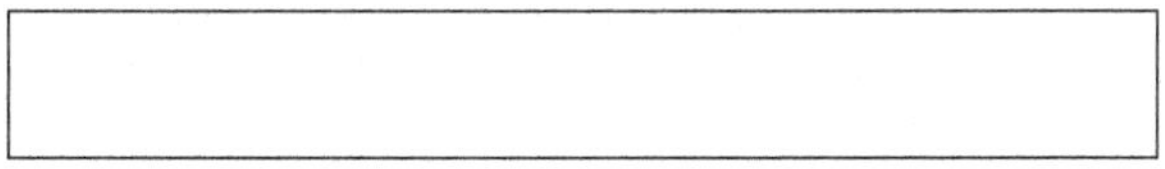

**Student response:**

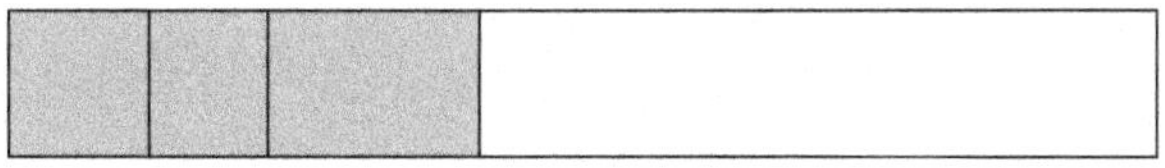

[Bar diagram divided into parts with 3 shaded sections]

I shaded 3 of the 4 parts to represent that Jorge walked $\frac{3}{4}$ of the way home.

| Describe the Error | Rework the Problem |
|---|---|
| | |

Figure 2.11: Grade 3 error analysis task.

**Error Analysis Task: Middle School**

**Directions:** Read the task and the student work. With your partner, identify the error in the student work and describe why it is an error. Rework the problem correctly.

**Task:** Evaluate $8 - (2^2 - 6)$.

| Student response: |
|---|
| $8 - (2^2 - 6)$ <br> $8 - (4 - 6)$ <br> $8 - (2)$ <br> $\boxed{6}$ |

| Describe the Error | Rework the Problem |
|---|---|
|  |  |

Figure 2.12: Middle school error analysis task.

---

**Error Analysis Task: High School**

**Directions:** Read the task and the student work. With your partner, identify the error in the student work and describe why it is an error. Rework the problem correctly.

**Task:** Find the center and radius of the circle given by the equation: $x^2 - 10x + y^2 + 6y - 30 = 0$

| Student response: |
|---|
| $x^2 - 10x + y^2 + 6y - 30 = 0$ <br> $x^2 - 10x + 20 + y^2 + 6y + 12 = 30 + 20 + 12$ <br> $(x - 5)^2 + (y + 3)^2 = 62$ <br><br> The center of the circle is at $(-5, -3)$ and the radius is $\sqrt{62}$. |

| Describe the Error | Rework the Problem |
|---|---|
|  |  |

Figure 2.13: High school error analysis task.

## Conclusion

Classroom culture contributes to student learning. Consider what students see in the physical space when they walk into your classroom and how they will become a learning community. Students need to feel a sense of belonging and that their thoughts are valued. Ask yourself, "How can I grow students' mathematical agency, identity, and positive self-efficacy through Tier 1 and Tier 2 learning experiences in this classroom?"

Look around the room at the organization of desks and consider how students are expected to share their thinking with their peers and access materials daily. Determine if the classroom norms for student learning are clear and accessible. When students are in a learning community that feels safe and acknowledges who they are, students are willing to take more risks, ask questions, and embrace mistakes, which helps them grow mathematically.

## Questions to Consider for Next Steps

As a collaborative team, use the following questions to reflect on your current practices and determine any next steps related to building a community of learners.

1. What current norms do you use to guide student-to-student interactions?
2. How are students included in creating classroom expectations?
3. How do you work to grow students' mathematics identities, mathematics agency, and self-efficacy?
4. How do students know the power of *not yet*?
5. How is the classroom set up to facilitate student-to-student discourse and group work?
6. What message is your physical space sending to students?
7. What tools do students use to share their thinking, and how can they quickly access these tools?
8. How can students learn from mistakes?

# PART 2

# Mathematics Foundations

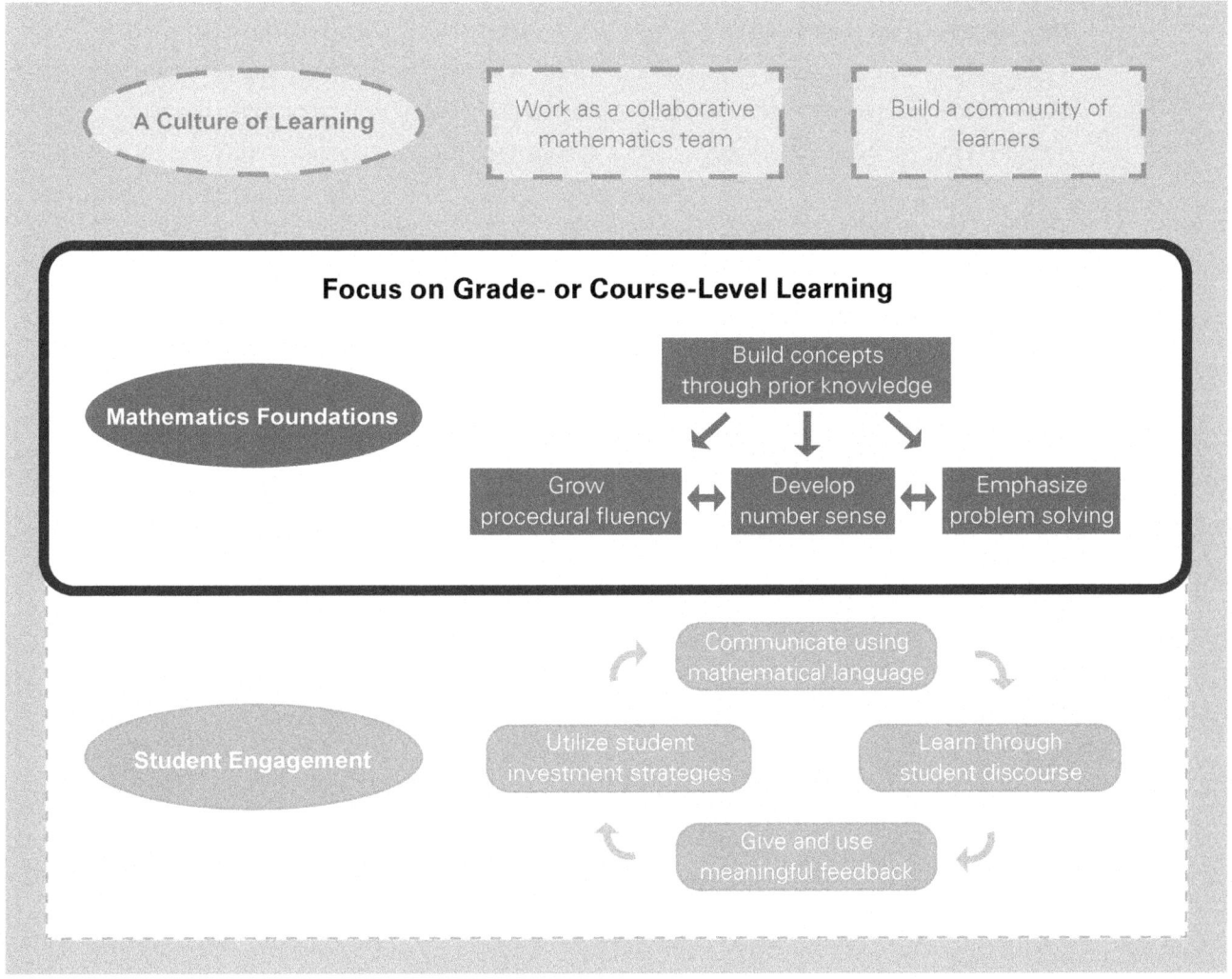

> Sarah was coaching a third-grade team in mathematics on the professional development day prior to Thanksgiving, and teachers came in very anxious and worried. She asked what was troubling them. They shared that they were behind in their pacing. Sarah asked, "How far behind?" Team members responded they were still in the first unit for the year because not every student was proficient with the content, and they did not want to move on to the next unit until every student was ready. However, they were supposed to be starting the fourth unit.
>
> The team also shared that their curriculum materials had very long lessons, and they needed to complete each one with fidelity, so they could not go as quickly as the district pacing guide suggested. They spent almost a week on each lesson. When Sarah and the team analyzed the mathematics learning in the first unit, they found it was primarily a review of second-grade standards. While the third-grade team's intentions were admirable—wanting every student to learn before moving on—there were many students who had learned the content quickly and were now relearning the same targets over and over.
>
> Students in third grade now only had about six months to learn what they should have had ten months (or more) to learn. Sarah brought in the fourth-grade team to work with the third-grade team to help clarify the mathematics essential standards most critical for students to learn in third grade. She wanted the fourth-grade team to be part of the solution and recognize what they could expect third graders to learn before entering fourth grade the next year.

# CHAPTER 3

# Teach Grade- or Course-Level Content

> If a student's core instruction is focused on below-grade-level standards,
> then he or she will learn well below grade level.
>
> —Austin Buffum, Mike Mattos, and Janet Malone

Like the third-grade team Sarah worked with (see story on page 42), it is not uncommon in some schools for mathematics teams to spend one or two months at the start of the year reviewing what students should have learned the previous year in an effort to prepare them for grade-level content. Sometimes, mathematics teams might also track students into leveled groups for core instruction, which means some students fall further behind because they are taught mathematics below grade level, while others move forward in grade-level content. Whether reviewing previous grade-level standards for a long time or tracking students, both actions prevent students from learning grade-level standards daily, which is critical to accelerating student learning to grade or course level and beyond.

Instead, Tier 1 and Tier 2 instruction needs to focus on grade- or course-level essential standards from the beginning of the school year with earlier content woven into units and lessons when it is directly connected to the grade-level standards students are learning.

## The *Why* and the *What* of Teaching Grade- and Course-Level Content

Each grade or course has specific mathematics state standards students must learn. In 1989, NCTM clarified the mathematics students should learn in each grade or course with the first set of standards in their book *Curriculum and Evaluation Standards for School Mathematics*. Since then, standards have continued to be revised by NCTM (2000, 2006), NGA & CCSSO (2010), and individual states. Accelerating student learning through Tier 1 and Tier 2 instruction necessitates knowing the mathematics state standards students need to learn in each grade or course and building shared knowledge of the learning expectations for each standard.

Many educational researchers and experts recognize the need to focus the mathematics standards in each grade level or course to what is essential for every student to learn (Ainsworth, 2004; DuFour et al., 2024; Kramer & Schuhl, 2023; Mattos et al., 2025; Reeves, 2002; Wiggins & McTighe, 2011). Regardless of state or province, there are still too many mathematics standards to teach in any given grade or course. Therefore, teams need to determine which mathematics standards are most important for students to learn. Sometimes, the essential standards are chosen by individual teams—other times they are chosen at the district level with representative teachers from each school in the district. The selected essential mathematics standards are those you and your collaborative team work to ensure, at a minimum, every student will learn at grade level to be successful in the next grade level or course.

After the COVID-19 pandemic, NCTM, NCSM, and Association of State Supervisors of Mathematics (ASSM) wrote a joint paper addressing the urgency of students' continued mathematics learning titled *Continuing the Journey: Mathematics Learning 2021 and Beyond* (NCTM et al., 2021). The paper addresses three key areas for equitable access to high-quality teaching and learning: (1) grade-level content, (2) equitable, effective teaching practices, and (3) advocacy. When addressing grade- or course-level content students need to learn, the three national mathematics organizations make the following recommendation:

> Regardless of what has come before, on-grade-level mathematics content must be the focus of our work with students. This includes on-grade-level mathematics content for PreK–8 and on-course-level mathematics content for high school. Particularly for the coming school year, mathematics educators must begin with building coherence through grade level content and shifting from remediation to supporting and scaffolding to facilitate student understanding. (p. 2)

Tracking students into leveled groups for initial learning is also addressed with a recommendation to eliminate the practice (NCTM et al., 2021). Prior to that recommendation, NCTM (2014) stated, "The practice of isolating low-achieving students in low-level or slower-paced mathematics groups should be eliminated" (p. 63). While students are grouped by targeted and specific need for intervention, core instruction is at grade or course level for all students.

Mattos and colleagues (2025) write that collaborative teams own quality Tier 1 instruction and targeted and specific Tier 2 interventions and extensions. Tier 1 and Tier 2 focus on essential standards. Mattos and colleagues (2025) emphasize that students will never learn grade-level standards if they do not receive access to grade-level learning. Thus, using tracking is not productive for student learning. Students need to be heterogeneously mixed for core instruction and grouped for a targeted and specific intervention, when needed.

Grade- or course-level essential standards are the focus for both Tier 1 and Tier 2 instruction. However, how these essential standards are taught and learned may differ between Tier 1 and Tier 2. In fact, Tier 2 learning is often more targeted and includes alternate ways of learning an essential standard than those used for Tier 1 instruction. Table 3.1 clarifies how to address grade- or course-level learning in both tiers.

Teaching essential standards to grade or course level (and above) requires you and your team to clarify essential standards and the learning progressions leading to each essential standard. Additionally, consider how you and your team will engage in conversations about instructional strategies and models to use, as well as ideas for differentiation and focused interventions for each unit of study. Ensure students at least learn the essential standards to grade or course level so they move on with critical learning in place.

Table 3.1: Teach Grade- or Course-Level Standards in Tier 1 and Tier 2

| Teach Grade- or Course-Level Standards in Tier 1 and Tier 2 ||
| --- | --- |
| <ul><li>Identify and focus instruction on essential standards or identified prior knowledge standards.</li><li>Teach rigorous mathematics, balancing conceptual understanding, application, and procedural fluency.</li><li>Utilize concrete objects and tools, pictures, and numbers and symbols to make sense of the mathematics skills being learned, and connect mathematical representations.</li></ul> ||
| **Teach Grade- or Course-Level Standards in Tier 1** | **Teach Grade- or Course-Level Standards in Tier 2** |
| <ul><li>Identify any important-to-know standards students should learn in a unit in addition to essential standards.</li><li>Create opportunities for students to connect prior learning to new grade- or course-level learning during instruction.</li><li>Differentiate and provide scaffolded supports to the essential standard being taught in each lesson.</li><li>Use mathematics strategies and models agreed on by the team when teaching.</li><li>Use and differentiate high-level-cognitive-demand tasks aligned to a grade- or course-level standard during lessons.</li></ul> | <ul><li>Focus targeted interventions to essential standards or prior knowledge content along a learning progression to the essential standard from earlier in the year or last year.</li><li>Emphasize conceptual understanding of the essential standards.</li><li>Use additional strategies and models not yet taught, and connect them to strategies and models used in Tier 1.</li><li>Use leveled tasks to differentiate learning to an essential standard.</li></ul> |

## The *How* of Teaching Grade- or Course-Level Content

Unfortunately, there is just not enough time in the school year to promise every student will learn every grade- or course-level standard. If teachers try to teach *all* standards, students are simply exposed to the standards and do not have the time to deeply learn and reason through any of them (Ainsworth, 2004; Marzano, 2003, 2010; Reeves, 2002).

### Essential Standards

To focus on essential standards means your team must identify which standards are considered essential. There are several protocols for determining these. Education authors Douglas B. Reeves (2002) and Larry Ainsworth (2013) set criteria for selecting essential standards that are still used today. Thomas W. Many and colleagues (2022) then captured the criteria using the acronym *REAL*.

1. **Readiness:** Is this standard needed for success in this subject area next year?
2. **Endurance:** Will students use this standard in their future?
3. **Assessment:** Does the blueprint for the state assessment show that this standard is important?
4. **Leverage:** Is this standard also used in other subject areas?

In *Acceleration for All*, authors Sharon V. Kramer and Sarah Schuhl (2023) ask, "Which eight to twelve standards would you design interventions for first if students are not yet learning any of the standards?" to distinguish the most essential (may also be called priority, power, promise, or critical) standards (p. 38). Ideally, your team determines these standards, as they are the focus of your Tier 1 and Tier 2 instruction (as shown in the RTI at Work pyramid in figure 1.2, page 12). Sometimes, districts create committees of teachers to determine the essential standards, in which case it becomes very important for your team to build a shared understanding of the rigor (conceptual understanding, application, and procedural fluency) associated with each standard.

The *Mathematics in a PLC at Work: Every Student Can Learn* series classifies standards as follows.

- **Essential standards:** Eight to twelve standards every student will learn to grade or course level or above and the focus for common mid-unit assessments and any team interventions
- **Important-to-know standards:** Standards that are also taught with essential standards in units and assessed on the common end-of-unit assessment
- **Nice-to-know standards:** Standards not taught and assessed but may be used for extension activities

Tier 1 instruction focuses first on essential standards and then addresses important-to-know standards. Tier 2 instruction solely focuses on essential standards—the standards all students will learn to grade or course level or above.

Teams create a list of essential standards and when proficiency is expected for each one. The *Mathematics Unit Planning in a PLC at Work* series provides directions for creating a year-at-a-glance proficiency map showing when students are expected to learn essential and important-to-know standards throughout the school year. It also offers directions for unwrapping (unpacking, deconstructing, or making sense of) the essential standards. Visit **go.SolutionTree.com/MathematicsatWork** to download directions and examples.

Teams unwrap standards one at a time as they prepare to teach each one. Part of the unwrapping process involves determining the skills within each standard and then creating a learning progression used to identify the order in which the unwrapped targets will be taught and learned. Such a progression guides Tier 1 instruction, informs common assessments, and helps determine targeted and specific Tier 2 interventions.

Students should experience grade- or course-level learning every day in their lesson (Kramer & Schuhl, 2023; Mattos et al., 2025; NCTM, 2014; NCTM et al., 2021; Toncheff et al., 2024). Creating connections from prior knowledge to daily-lesson learning is one of the six criteria for highly effective lesson design in *Mathematics Instruction and Tasks in a PLC at Work, Second Edition* (Toncheff et al., 2024). During Tier 1 instruction, determine how much instructional time will be devoted to prior knowledge and how much to grade-level standards using data from team common mid-unit or end-of-unit assessments or from a pretest asking questions related to prior knowledge standards.

In Tier 2, students focus on a specific target in an essential standard or prior knowledge standard along the learning progression to the standard. The mathematics coherence map at www.achievethecore.org can support your team in identifying the direct prior knowledge standards so you re-engage students in targeted concepts and skills instead of trying to reteach everything. Chapter 4 (page 57) explores specific instructional routines to use when connecting grade-level learning to prior knowledge.

Figure 3.1 shows examples of how a learning target from an essential standard might be addressed in Tier 1 and Tier 2.

Use the protocol in figure 3.2 (page 49) to determine how to address grade- or course-level learning every day when planning your lessons.

It is possible that more of a lesson may need to be devoted to prior knowledge than grade- or course-level content based on student readiness (using data from previous common assessments or a pretest). However, every lesson still has some grade- or course-level content—over time, lessons grow to primarily address grade- or course-level learning with a nod to prior knowledge using prior knowledge routines, which chapter 4 (page 57) explores in more detail.

Student work from common assessments reveals which targets to use in Tier 2 instruction (see *Mathematics Assessment and Intervention in a PLC at Work, Second Edition,* Schuhl et al., 2023). To determine targeted and specific interventions, sort student work and identify as a team what the proficient students showed in their work. Then, look at the students with a partial understanding. What, as a trend, did they understand, and what is needed to move them to proficiency based on the work proficient students demonstrated? Teams will begin to see they do not need to reteach everything, but rather focus on skills students have not yet learned. Finally, determine any trends among students who have minimal proficiency and determine what is needed to move them to partial proficiency. Your targeted intervention may be part of the essential standard or a prior knowledge standard. Work as a team to determine the models and strategies used when teaching each student group.

### High-Level-Cognitive-Demand Tasks

Students learning grade- or course-level content every day includes engagement in high-level-cognitive-demand tasks. A mathematical task can be multiple questions or just one question focused on a specific mathematical concept. Using the work of education professors Margaret S. Smith and Mary K. Stein (2011), tasks are described as low-level-cognitive-demand tasks (more procedural in nature) and high-level-cognitive-demand tasks (higher-order reasoning with possible options for multiple models or strategies when solving; see appendix B, page 179, for the Cognitive-Demand-Level Task Analysis Guide). A balance of both is needed during instruction, as described in *Mathematics Instruction and Tasks in a PLC at Work, Second Edition* (Toncheff et al., 2024).

There is a connection between students engaging in high-level-cognitive-demand tasks and overall increases in student achievement (Hattie, 2012; Resnick & Zurawsky, 2006; Smith, Steele, & Raith, 2017). Your choice of tasks dictates what students will learn and will either limit or extend that learning. High-level-cognitive-demand tasks work to enrich, grow, and meaningfully challenge mathematical reasoning.

In addition to increasing students' mathematical reasoning through high-level-cognitive-demand tasks, such tasks provide opportunities to differentiate and promote student learning in groups using discourse (see chapter 9, page 129). Educator and author Carol A. Tomlinson (2016) focuses much of her educational work on differentiation:

> Differentiation is a model designed to guide teaching that provides equity of access to excellence for every student. To that end, the teacher in a differentiated classroom believes in the capacity of every student to succeed, works from curriculum that requires every student to grapple with the essential understandings or principles of a discipline and to be a thinker and problem solver in the context of that curriculum, scaffolds the next steps of every learner in a progression toward and beyond critical learning goals, and creates a classroom that actively supports the growth of each of its members. (p. 2)

Teach Grade- or Course-Level Content     47

| Learning Target | Tier 1 Content for Prior Knowledge Routines | Tier 1 Content at Grade or Course Level | Possible Tier 2 Content After a Common, Mid-Unit Assessment Tied to the Learning Target |
|---|---|---|---|
| **Kindergarten** Solve addition and subtraction word problems within 10 by using objects or drawings to represent the problem. | Launch the lesson with students counting 6 cubes. Students will organize the count using a ten frame in a way that makes sense to them. Ask the following. <br>• How do you know it is 6? <br>• Show one more than 6 with your cubes. | • Give students a story mat like the picture of this fishpond. <br><br>• Tell students they are going to solve problems about fish that they can act out with their cubes. <br>Example fish story problems: <br>1. There are 3 fish in the pond. 4 more fish swim in and join the others. How many fish are in the pond now? <br>2. I see 4 blue fish and 2 yellow fish. How many fish do I see all together? <br>3. There are 8 fish swimming as a group, but 3 fish swam away. How many fish are still swimming together? | Allow students to solve problems within a real context, like eating cereal pieces or fruit snacks using a ten frame. The frame will help students keep track of the count, practice subitizing, and model one more and one less. The example problems in the Subtraction Cereal Goldfish Lesson (see bit.ly/3TauYam) relate directly to the tool students are using to count. |
| **Grade 2** Solve one-step word problems using addition and subtraction within 100. | Launch the lesson with a one-step word problem within 20 (first-grade standard). <br>*Molly ran 12 miles. Camden ran on the same day as Molly. Together they ran 19 miles. How many miles did Camden run?* | • Give students one word problem on a strip of paper or on a slide. Ask students to use the three reads protocol (see chapter 6, page 91) to make sense of the problem. Students solve the problem on a whiteboard. Compare models and strategies used. <br>• Repeat with new word problems, one at a time. After doing the first two as a whole class, have teams work independently, check their solutions with the teacher, and then work on new practice problems. | • Give students linking cubes to model and solve the word problem. Once they reason to a solution, have them draw a picture to show their thinking and then write an equation. Repeat. <br>• Start with word problems within 20 without regrouping and then with regrouping before moving to grade-level word problems. |
| **Grade 4** Find all factor pairs for a whole number in the range 1–100. | Launch the lesson by reviewing arrays linked to multiplication. <br>*Gene is organizing his shoe rack. He has 14 pairs of shoes and wants to put the same number of shoes in two rows. Is this possible? Why or why not?* | • During the lesson, start by using arrays to show the factor pairs for a given whole number. Ask a question like the following. <br>*The local high school marching band is participating in a parade. The band marches in an array. How many ways can the 48-person band be arranged?* <br>• Have students share their solutions with the class to generate all the factor pairs. Repeat with a similar question, using the number 36. | Using physical objects, such as beans, counters, linking cubes, or drawings, have students create an array for each possible formation. |

**Figure 3.1: Examples of how to teach content in Tier 1 and Tier 2 for daily learning targets.**

continued ▶

| Learning Target | Tier 1 Content for Prior Knowledge Routines | Tier 1 Content at Grade or Course Level | Possible Tier 2 Content After a Common, Mid-Unit Assessment Tied to the Learning Target |
|---|---|---|---|
| **Grade 7**<br><br>Solve two-step equations. | Launch the lesson with two one-step equations (sixth-grade standard), such as the following.<br><br>$x + 12 = 20$<br><br>$45 \div a = 5$<br><br>Students work in pairs to solve each equation. Notice if they are using number sense, an algorithm, or another strategy to solve it. | • Start the grade-level lesson with a word problem in which students solve a two-step equation using positive whole numbers.<br><br>*Maya made 42 cookies to take to school. She gave 14 cookies to her choir group and gave some more cookies to her friends at lunch. At the end of the day, she still had 15 cookies. How many cookies did Maya give to her friends at lunch?*<br><br>• Use the context to build understanding and show student thinking with equations.<br><br>• Repeat with a word problem involving at least one fraction or decimal.<br><br>• Give students a two-step equation to solve, such as $2x + 20 = 30$. Students solve it in pairs on a whiteboard without any prior instruction. Compare ideas for how to find the value of $x$.<br><br>• Repeat and then bring students together to discuss strategies to solve two-step equations. Students practice additional examples in pairs (some with fractions or decimals). | • Use manipulatives to create a one-step or two-step equation (depending on entry point for students). Draw a picture to represent the problem and then write the equation and show algebraic thinking to solve.<br><br>• Consider using a balance scale with blocks as manipulatives to show that the equation must stay balanced (for example, subtract the same value from both sides of the equation).<br><br>• Consider giving students the answer to substitute into the equation so they understand what it means to solve for a variable. |
| **Geometry**<br><br>Find the area of special quadrilaterals and triangles on a coordinate plane. | • Launch the lesson by giving students a right triangle, rectangle, and parallelogram with dimensions and ask students to find the area of each one.<br><br>• Give students two points on a coordinate plane (not directly horizontal or vertical from one another) and ask them to find the distance between the points (students might create right triangles or use the distance formula). | • Start the course-level part of the lesson by having students graph a triangle on a coordinate plane and find its area. Use coordinates like (−4, 2), (6, 2) and (1, −3) so the triangle dimensions needed are along horizontal and vertical lines. Repeat with a parallelogram.<br><br>• Give students ordered pairs for triangles and special quadrilaterals with key dimensions not along horizontal or vertical lines so students cannot just count to see the side lengths needed for the area. Repeat. Give students the answers to about four problems while they work in groups to graph and find the areas. Students must show why the answer given is correct. | • Address one or two of the following prior knowledge standards.<br>  ▸ Area of rectangles (includes squares)<br>  ▸ Area of triangles<br>  ▸ Area of parallelograms that are not rectangles (includes rhombus)<br>  ▸ Plot points on a coordinate plane to create and name a shape<br><br>• Find the area of a shape on a coordinate plane that has horizontal and vertical key dimensions. Provide a drawn shape; then have students create the shape and find its area. |

**Directions:** Answer the following questions to determine content and strategies to include in Tier 1 when teaching grade- or course-level essential standards.

1. What is today's daily learning target to focus the lesson?

2. What will students have to do at grade or course level to demonstrate understanding of the daily learning target?

3. What types of tasks should you use to teach the daily learning target to grade or course level?

4. What prior knowledge do students need that connects to the lesson's learning target?

5. What task (or tasks) can you use to engage students in remembering necessary prior knowledge (without needing any additional instruction to complete it)?

**Figure 3.2: Protocol to determine how to teach grade- or course-level learning in Tier 1 lessons.**

*Visit* **go.SolutionTree.com/MathematicsatWork** *for a free reproducible version of this figure.*

Differentiate mathematics by scaffolding and extending high-level-cognitive-demand tasks, most often used in Tier 1 core instruction. Work with your team to identify such a task, and then discuss the different models or strategies students might use to make sense of and solve the task. After, work with your team to address the following questions (adapted from Toncheff et al., 2024).

- What different models or strategies might students use to solve the task?
- What types of misconceptions do you anticipate students will struggle with during the task?
- What scaffolding questions can you ask or what tools might you give students to help guide their work?
- How might feedback for the student solution pathway or explanation be provided during the lesson (teacher to student or student to student)?
- How can the task be extended, or learning continued, if students finish early?

When answering these questions about a high-level-cognitive-demand task, consider how to provide scaffolded supports without lowering the cognitive demand by telling students *how* to solve the task. Figures 3.3–3.7 (pages 50–52) show a protocol for asking questions to plan for differentiation across grade levels.

Planning scaffolded questions or tools to provide to student groups when solving a high-level-cognitive-demand task differentiates entry points into the task while keeping the complexity of reasoning high. Creating a plan to extend learning for students who finish early is another way to differentiate learning and ensure students are engaging with mathematics during the full block of time or class period. Another option for differentiation includes using varied tasks to reach the same target or standard, as described in the following section.

### Leveled Tasks

Differentiation also involves using different tasks or activities aligned to the same standard. Rather than give every student the same task or activity, consider using tasks along a learning progression to the grade-level standard that produces similar mathematical reasoning. Leveled tasks are often used in Tier 1, but they are also an effective Tier 2 strategy designed to accelerate student learning. Differentiation, at times, involves creating varied assignments that engage students in appropriately challenging tasks—not too easy and not too difficult (Strickland, 2007).

| **Grade PreK** |  |
|---|---|
| **Standard:** Count to 10 and identify shapes. | |
| **Task:** Match pattern blocks to each shape that makes the picture. Count the triangles and say how many triangles there are in all. 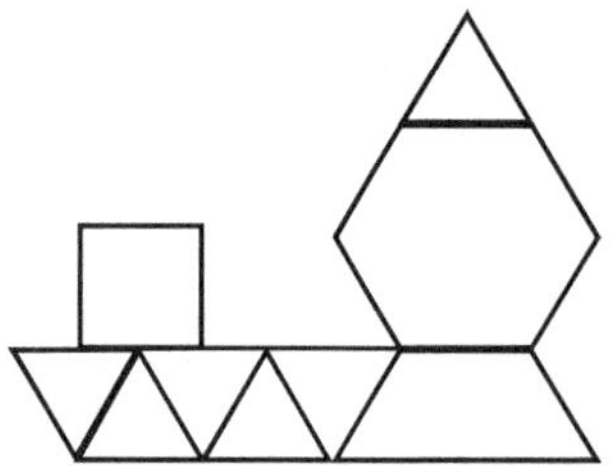 | |
| What different models or strategies might students use to solve the task? | |
| What types of misconceptions do you anticipate students will struggle with during the task? | |
| What scaffolding questions can you ask or what tools might you provide students to help guide their work? | |
| How might feedback for the student solution pathway or explanation be provided during the lesson (teacher to student or student to student)? | |
| How can the task be extended, or learning continued, if students finish early? | |

*Source: Adapted from Toncheff et al., 2024.*

**Figure 3.3: Differentiation plan for a high-level-cognitive-demand task—preK.**

*Visit **go.SolutionTree.com/MathematicsatWork** for a free blank reproducible version of this figure.*

| **Grade 2** |  |
|---|---|
| **Standard:** Solve word problems using addition and subtraction within 100. | |
| **Task:** Jose ran 18 laps during a school fun run. His friend Carlos ran 3 fewer laps. How many laps did Jose and Carlos run altogether? | |
| What different models or strategies might students use to solve the task? | |
| What types of misconceptions do you anticipate students will struggle with during the task? | |
| What scaffolding questions can you ask or what tools might you provide students to help guide their work? | |
| How might feedback for the student solution pathway or explanation be provided during the lesson (teacher to student or student to student)? | |
| How can the task be extended, or learning continued, if students finish early? | |

*Source: Adapted from Toncheff et al., 2024.*

**Figure 3.4: Differentiation plan for a high-level-cognitive-demand task—grade 2.**

| **Grade 3** |  |
|---|---|
| **Standard:** Solve word problems involving multiplication and division within 100. | |
| **Task:** During a fundraiser, 6 of the students sold a total of 48 cakes. How many cakes did each student sell if they all sold the same number of cakes? | |
| What different models or strategies might students use to solve the task? | |
| What types of misconceptions do you anticipate students will struggle with during the task? | |
| What scaffolding questions can you ask or what tools might you provide students to help guide their work? | |
| How might feedback for the student solution pathway or explanation be provided during the lesson (teacher to student or student to student)? | |
| How can the task be extended, or learning continue, if students finish early? | |

*Source: Adapted from Toncheff et al., 2024.*

**Figure 3.5: Differentiation plan for a high-level-cognitive-demand task—grade 3.**

| **Grade 6** |  |
|---|---|
| **Standard:** Solve word problems involving operations with decimals. | |
| **Task:**<br>A tortoise travels 1.44 miles in 4.5 hours.<br>    a. If the tortoise moves at a constant rate, how many miles per hour is it traveling?<br>    b. How long will it take the tortoise to travel 5.12 miles? | |
| What different models or strategies might students use to solve the task? | |
| What types of misconceptions do you anticipate students will struggle with during the task? | |
| What scaffolding questions can you ask or what tools might you provide students to help guide their work? | |
| How might feedback for the student solution pathway or explanation be provided during the lesson (teacher to student or student to student)? | |
| How can the task be extended, or learning continued, if students finish early? | |

*Source: Adapted from Toncheff et al., 2024.*

**Figure 3.6: Differentiation plan for a high-level-cognitive-demand task—grade 6.**

| Integrated Mathematics 2 |
|---|
| **Standard:** Interpret key features of quadratic functions in context. |
| **Task:**<br>A batter hits a pitched baseball. After it is hit, the height $h$ (in feet) of the ball at time $t$ (in second) is modeled by $h = -16t^2 + 80t + 3$.<br>  a. How high off the ground was the baseball when it was hit?<br>  b. How long did it take for the baseball to land in the outfield?<br>  c. What was the maximum height of the baseball and how long did it take to reach that height once the baseball was hit? |

| | |
|---|---|
| What different models or strategies might students use to solve the task? | |
| What types of misconceptions do you anticipate students will struggle with during the task? | |
| What scaffolding questions can you ask or what tools might you provide students to help guide their work? | |
| How might feedback for the student solution pathway or explanation be provided during the lesson (teacher to student or student to student)? | |
| How can the task be extended, or learning continued, if students finish early? | |

*Source: Adapted from Toncheff et al., 2024.*

**Figure 3.7: Differentiation plan for a high-level-cognitive-demand task—integrated mathematics 2.**

In kindergarten, students compare numbers 0 through 9 written as numerals. Some students are still making sense of a number's value and may need to use objects to show each numeral to compare (developing task). Later, students may be able to draw a picture of the numerals to compare (approaching task). Eventually, students at grade level compare two written numerals (grade-level task). When used in Tier 1 instruction, an extension task may be needed for some students. Students could compare larger numbers or create a comparison of multiple numbers (extension task). Figure 3.8 shares an example of leveled tasks for kindergarten using a teacher team planning template.

When teaching students in second grade how to add and subtract within 100 or 1,000, some students may still be working to add and subtract within twenty. A leveled task in Tier 2 might ask students to solve word problems using addition and subtraction within 20 (developing task). Over time, the tasks are altered to add and subtract within 50 and then 100 (approaching and grade-level task). In every task, students are reading word problems, adding or subtracting with and without regrouping, and matching the mathematical reasoning required in the grade-level standard. If used as part of differentiation in Tier 1, leveled tasks also include an extension task.

| Developing Task | John has 2 toy cars. Nancy has 4 toy cars. Use cubes to show how many cars John has and how many cars Nancy has. Who has more toy cars? |
|---|---|
| Approaching Task | Paul has 7 toy cars. Maria has 3 toy cars. Draw a picture to show how many toy cars Paul has and how many toy cars Maria has. Who has more toy cars? |
| Grade-Level Task | Gabe has 9 toy cars. Alexa has 7 toy cars. Who has more toy cars? |
| Extension Task | Sammy has 6 toy cars. Jamil has more toy cars than Sammy. How many toy cars might Jamil have? Is there more than one answer? Explain. |

**Figure 3.8: Kindergarten leveled-tasks example.**

*Visit **go.SolutionTree.com/MathematicsatWork** for a free blank reproducible version of this figure.*

Figure 3.9 shares an example of leveled tasks using a teacher team planning template. Do not include the type of task when giving students tasks to choose from, only the task question.

In fifth grade, students might be learning to multiply fractions. Give word problems or algebraic tasks to students that require multiplication of a unit fraction by a whole number (fourth-grade developing task), a fraction by a whole number (approaching task), and eventually a fraction by a fraction (grade-level task). Use models to demonstrate the concept of multiplying fractions. Again, the mathematical reasoning is consistent, but the numbers used differ, as shown in the grade 5 leveled tasks in figure 3.10.

In seventh grade, students might be learning to solve proportions. Use word problems or algebraic tasks to find ratios (developing task), find equivalent ratios (approaching task), and then solve proportions (grade-level task). Students may mathematically reason using a ratio table, graph, number lines, double number lines, equivalent ratios, or a comparison of two ratios, or they can solve a proportion with a missing value. The mathematical reasoning of ratios to proportions is consistent through the leveled tasks, as shown in figure 3.11.

In geometry, students who are completing coordinate proofs to prove that a given shape is a special quadrilateral might first be given coordinates for a rectangle with all sides horizontal and vertical (developing task).

| | |
|---|---|
| Developing Task | Carl is making cookies. He wants to take 15 cookies to soccer practice. Carl made 7 cookies in his first batch and 3 cookies in his second batch. How many more cookies does Carl need to make? |
| Approaching Task | Theo is making cookies. He wants to take 52 cookies to school. Theo makes 18 cookies in the morning and 24 cookies in the afternoon. How many more cookies does Theo need to make? |
| Grade-Level Task | Lea is making cookies. She wants to take 75 cookies to school for a party. She has made 52 cookies, but her friend ate 4 of them. How many more cookies does Lea need to make? |
| Extension Task | Sam is in charge of the cookie table at a bake sale. He needs to sell 120 cookies. Maggie gives Sam 38 cookies to sell, Charlie gives Sam 56 cookies to sell, and Juan gives Sam 19 cookies to sell. Does Sam have enough cookies to sell? Explain why or why not. |

**Figure 3.9: Grade 2 leveled-tasks example.**

| | |
|---|---|
| Developing Task | A container holds 3 pound of jellybeans. The container is $\frac{1}{4}$ full. How many pounds of jellybeans are in the container? |
| Approaching Task | A container holds 5 pounds of jellybeans. The container is $\frac{2}{3}$ full. How many pounds of jellybeans are in the container? |
| Grade-Level Task | A container holds $\frac{1}{2}$ pounds of jellybeans. The container is $\frac{3}{5}$ full. How many pounds of jellybeans are in the container? |
| Extension Task | Julie wants as many jellybeans as possible. She can choose from container A or container B. Container A holds $2\frac{1}{2}$ pounds of jellybeans and is $\frac{3}{4}$ full. Container B holds 3 pounds of jellybeans and is $\frac{2}{5}$ full. Which container should Julie choose and why? |

**Figure 3.10: Grade 5 leveled-tasks example.**

| | |
|---|---|
| Developing Task | A train travels 250 miles at a constant speed in 4 hours. What is the rate the train travels in miles per hour? |
| Approaching Task | Train A travels 250 miles at a constant speed in 4 hours. Train B travels 434 miles in 7 hours. Which train is faster? Explain your answer. |
| Grade-Level Task | A train travels 368 miles at a constant speed in 5 hours. How many miles will the train travel in 7 hours at the same constant speed? |
| Extension Task | Train A travels 400 miles at a constant speed in 8 hours. Train B travels at a rate of 82 miles per hour for 8 hours. Train C travels 75 miles every 5 hours. If the trains race for 500 miles, which train will win and why? |

**Figure 3.11: Grade 7 leveled-tasks example.**

Next, give students a quadrilateral with one pair of opposite sides that are horizontal or vertical (approaching task). At grade level, give students quadrilaterals without any horizontal or vertical sides (grade-level task). To extend the task, you may give students variable ordered pairs (extension task). Figure 3.12 shows the leveled-task progression.

In Tier 2 interventions, leveled tasks provide an opportunity to re-engage students in learning prior knowledge standards and connect that learning to grade- or course-level learning. The leveled tasks do not stop at the level learned in prior grades or courses, rather they are used as a catalyst for grade- or course-level learning by making connections and growing students' confidence as they graduate through the leveled tasks.

## Conclusion

To learn grade- or course-level standards, students need to engage in learning grade- or course-level mathematics every day during Tier 1 instruction. That learning can be scaffolded or differentiated using high-level-cognitive-demand tasks or grown through strategically chosen leveled tasks in Tier 2 learning experiences that grow to grade- or course-level expectations.

The types of tasks chosen to use in class matter, especially when accelerating student learning to grade level and beyond. With your team, identify the types of tasks to use in class, the prior knowledge leading to the grade- or course-level standard to teach in Tier 1 or address in Tier 2, and the varied ways to scaffold and extend learning consistent with the mathematical reasoning required in the standard.

## Questions to Consider for Next Steps

As a collaborative team, use the following questions to reflect on your current practices and determine any next steps related to teach grade- or course-level content every day.

1. What is a challenge of teaching grade- or course-level learning every day?

2. What grade- or course-level essential standards will students learn in your current or next unit?

| | |
|---|---|
| **Developing Task** | Graph and connect the following points.<br>$A(2, 4)$<br>$B(7, 4)$<br>$C(7, 8)$<br>$D(2, 8)$<br>Determine which type of special quadrilateral is graphed and prove why. |
| **Approaching Task** | Graph and connect the following points.<br>$A(-4, 1)$<br>$B(-2, 7)$<br>$C(9, 7)$<br>$D(7, 1)$<br>Determine which type of special quadrilateral is graphed and prove why. |
| **Grade-Level Task** | Graph and connect the following points.<br>$A(-4, 6)$<br>$B(-1, 5)$<br>$C(-3, -1)$<br>$D(-7, 0)$<br>Determine which type of special quadrilateral is graphed and prove why. |
| **Extension Task** | Graph and connect the following points.<br>$A(a, 0)$<br>$B(0, a)$<br>$C(-a, 0)$<br>$D(0, -a)$<br>Determine which type of special quadrilateral is graphed and prove why. |

**Figure 3.12: Geometry leveled-tasks example.**

3. How do you and your collaborative team determine the prior knowledge needed to learn the essential grade- or course-level standards?

4. How do you and your team utilize concrete objects and tools when teaching conceptual understanding of a standard in Tier 1 and Tier 2?

5. How do you and your team choose the high- and low-level-cognitive-demand tasks students need to solve to be grade- or course-level proficient?

6. How does your team differentiate using high-level-cognitive-demand tasks during Tier 1 instruction?

7. How does your team differentiate learning during Tier 2 interventions using leveled tasks requiring similar mathematical reasoning as grade- or course-level tasks?

> When Dr. Buck (Brian) was coaching a second-grade team, the team expressed frustration with figuring out how to support students who were still struggling to learn subtraction and regrouping. When questioned about which strategies they had already tried, the teachers said they relied mostly on providing more practice opportunities—sometimes in a whole-group setting and other times in small-group or one-on-one settings. However, even as students continued practicing, they were not showing any progress in their understanding. Dr. Buck began to question the teachers more about which parts or processes of subtraction with regrouping were causing the students' struggles. Almost in unison, the teachers responded, "Everything!"
>
> Dr. Buck paused the discussion for a moment and wrote an example problem on the board. He asked the team to list all the things students need to know or be able to do to successfully subtract with regrouping. The teachers began to compile a list of concepts, such as place-value knowledge, the relationship between tens and ones, the connection between addition and subtraction, and regrouping with tools, pictures, and written numbers. To help the second-grade teachers organize and formalize their thoughts, Dr. Buck explored a digital coherence map with the team that allowed the teachers to see mathematics standards students previously learned to support their students' understanding of subtraction with regrouping (see www.achievethecore.org). The teachers examined conceptual strategies, models and tools, and procedural expectations found in the previously learned standards.
>
> Dr. Buck then asked the teachers to think of one student still learning how to subtract with regrouping and, based on their classroom observations and work samples from that student, identify a potential previous standard from the coherence map that the student may need more support learning. The teachers started naming specific students and their areas of need from the previous standards.
>
> To gather more evidence and determine the specific needs of their students as a team, the team created a couple of problems to assess students' knowledge of prior skills so they could create a plan to re-engage any students still working to learn prior skills, before reconvening their work in subtraction with regrouping. The teachers left the meeting with a collective game plan to identify previous learnings, including tools and tasks, to use to accelerate student learning.

CHAPTER 4

# Connect to Prior Knowledge

> Effective mathematics teaching begins with a shared understanding among teachers of the mathematics that students are learning and how this mathematics develops along learning progressions.
>
> —NCTM, Principles to Action: Ensuring Success for All

When students are not able to access grade- or course-level content, teachers—like the second-grade teachers working with Dr. Buck (see story on page 56)—are often quick to reteach everything students should have learned in mathematics during previous grades or give more practice problems focused on procedural fluency. For example, when teaching ratios, a sixth-grade teacher might go back and teach single-digit addition and subtraction from the primary grades or give students worksheets with many problems to answer related to ratios. Teachers strategically place prior knowledge concepts in units as part of Tier 1 and Tier 2 instruction when it is a prerequisite to a grade-level essential standard. However, it's important to note that prior knowledge from several years before often becomes part of a Tier 3 intervention.

## The *Why* and the *What* of Connecting to Prior Knowledge

The structure of mathematics is founded on a small collection of fundamental concepts, generalizations, and properties rather than a larger set of individualized, isolated, and unrelated skills (Common Core Standards Writing Team, 2013). Students in primary and early intermediate grades often learn concepts with whole numbers first and then later extend and apply those concepts to more complex sets of numbers, such as fractions, decimals, rational numbers, and real numbers. This extension applies to models and tools, such as the number line. Students initially interact with the number line via whole numbers, and they eventually extend its use to include fractions, integers, rational numbers, and real numbers (Common Core Standards Writing Team, 2013; see chapter 5, page 69).

From the first set of mathematics standards to present-day mathematics standards in every state and province, there continues to be a more explicit focus on the coherence—the connections among and between—the mathematics content standards that identify what students should know and be able to do at each grade level (Battelle for Kids, 2013; Lemon & Hendrickson, 2023; NCTM, 2006). The 2010 Mathematics Common Core State Standards (NGA & CCSSO, 2010) and most standards in each state or province continue to be created with coherence in mind. While focusing on creating a more coherent set of learning standards, special attention was given to learning progressions or trajectories of the mathematics content, acknowledging the importance of sequencing content with attention to how students learn best. Consider how student learning of addition progresses through the grade levels with the standards in figure 4.1 (page 58).

Learning progressions (also referred to as learning trajectories or paths) describe a purposeful, strategic sequencing to best support each learner (NCTM, 2014). According to the Common Core Standards Writing Team (2013), three essential components are embedded within learning progressions: (1) a specific mathematical goal, (2) a path for students to reach that goal, and (3) the instructional decisions to support students moving along the path to the goal. Along each path, teacher teams develop systems to monitor student learning and communicate progress to the learner.

**Directions:** Read the addition learning progression of standards showing some of the vertical standards students need to learn to add. The addition progression shows standards over the course of several grade levels. Respond to the following questions with your team.

- How are the addition standards in the learning progression connected?
- What specific connections do you see from one grade to the next?
- How are students understanding the concept of addition, which can be applied to different sets of numbers, and even polygons, from one year to the next?

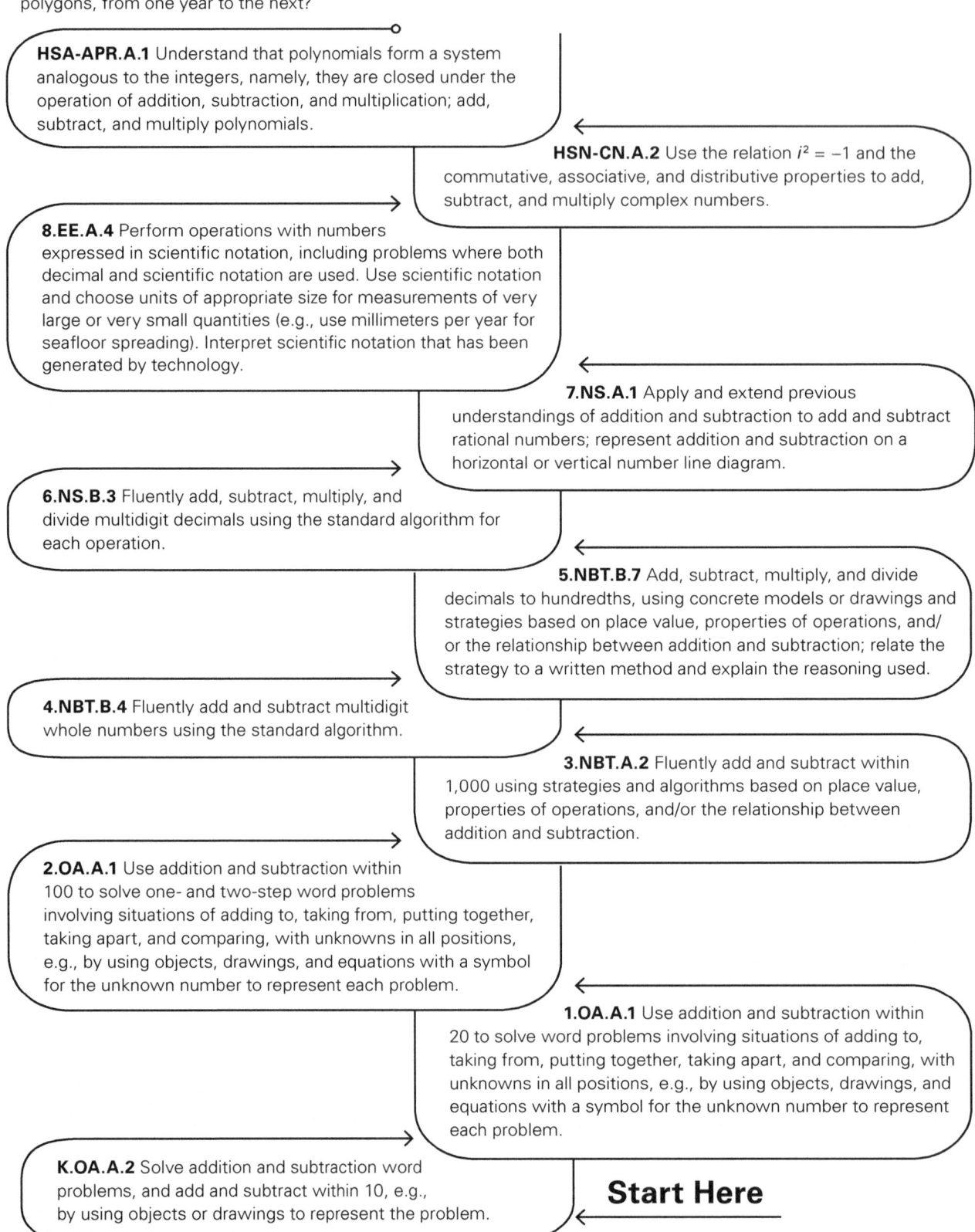

**HSA-APR.A.1** Understand that polynomials form a system analogous to the integers, namely, they are closed under the operation of addition, subtraction, and multiplication; add, subtract, and multiply polynomials.

**HSN-CN.A.2** Use the relation $i^2 = -1$ and the commutative, associative, and distributive properties to add, subtract, and multiply complex numbers.

**8.EE.A.4** Perform operations with numbers expressed in scientific notation, including problems where both decimal and scientific notation are used. Use scientific notation and choose units of appropriate size for measurements of very large or very small quantities (e.g., use millimeters per year for seafloor spreading). Interpret scientific notation that has been generated by technology.

**7.NS.A.1** Apply and extend previous understandings of addition and subtraction to add and subtract rational numbers; represent addition and subtraction on a horizontal or vertical number line diagram.

**6.NS.B.3** Fluently add, subtract, multiply, and divide multidigit decimals using the standard algorithm for each operation.

**5.NBT.B.7** Add, subtract, multiply, and divide decimals to hundredths, using concrete models or drawings and strategies based on place value, properties of operations, and/or the relationship between addition and subtraction; relate the strategy to a written method and explain the reasoning used.

**4.NBT.B.4** Fluently add and subtract multidigit whole numbers using the standard algorithm.

**3.NBT.A.2** Fluently add and subtract within 1,000 using strategies and algorithms based on place value, properties of operations, and/or the relationship between addition and subtraction.

**2.OA.A.1** Use addition and subtraction within 100 to solve one- and two-step word problems involving situations of adding to, taking from, putting together, taking apart, and comparing, with unknowns in all positions, e.g., by using objects, drawings, and equations with a symbol for the unknown number to represent each problem.

**1.OA.A.1** Use addition and subtraction within 20 to solve word problems involving situations of adding to, taking from, putting together, taking apart, and comparing, with unknowns in all positions, e.g., by using objects, drawings, and equations with a symbol for the unknown number to represent each problem.

**K.OA.A.2** Solve addition and subtraction word problems, and add and subtract within 10, e.g., by using objects or drawings to represent the problem.

**Start Here**

*Source for standards: NGA & CCSSO, 2010.*

**Figure 4.1: Example of an addition learning progression.**

Learning progressions support teachers when planning instruction, differentiating instruction for all students to meet grade-level expectations, scaffolding learning experiences, and monitoring each student's progress (Battelle for Kids, 2013).

In addition to planning and monitoring student learning, learning progressions allow teachers to look back at previously taught standards in earlier grades to identify the prior knowledge students need to access current grade-level standards. The standards along the progression are sometimes taught as part of Tier 1 instruction and differentiation and other times as part of Tier 2 targeted learning experiences. Identifying specific immediate prior knowledge standards to an essential standard is key to making meaningful connections to current learning for students.

Teachers address prior knowledge during both Tier 1 and Tier 2 instruction. The purpose for addressing prior knowledge varies slightly in each tier, as shown in table 4.1.

Student learning is accelerated to grade level (and beyond) when students learn the prior knowledge aligned to the grade-level essential standard. You and your team will address more immediate prior knowledge standards in Tier 1 and Tier 2 while including grade-level learning in each daily lesson (see chapter 3, page 43).

## The *How* of Connecting to Prior Knowledge

In mathematics, students' mastery of prior knowledge skills is foundational for current and future learning. As students make connections between what they already know and can do and new concepts, their learning becomes more coherent and meaningful (NCTM et al., 2021). To grow coherence, your team makes important decisions regarding when to provide support in strengthening prior knowledge and uses strategies to help students make connections to prior learning. The following strategies address prior knowledge and support student learning of grade- or course-level mathematics.

### Preassessments

Teachers use valuable instructional minutes when giving common mid-unit and end-of-unit assessments. As such, the data from each assessment should be used immediately afterward to continue student learning (for example, learning through revising work or re-engaging in a targeted intervention) and provide information for you and your team to learn about student thinking (Schuhl et al., 2024). At times, you and your team may decide to administer a preassessment to determine whether students have the prior knowledge needed to learn an upcoming unit's essential standards.

Table 4.1: Connect to Prior Knowledge in Tier 1 and Tier 2

| Connect to Prior Knowledge in Tier 1 and Tier 2 ||
|---|---|
| <ul><li>Utilize progressions as a collaborative team to identify prior knowledge standards students must learn to appropriately support current grade- or course-level learning.</li><li>Identify models and strategies from previous learning experiences to support current grade- or course-level learning.</li><li>Promote flexible thinking based on prior knowledge strategies.</li><li>Support students in connecting new concepts to previous learning through visuals.</li></ul> ||
| **Connect to Prior Knowledge in Tier 1** | **Connect to Prior Knowledge in Tier 2** |
| <ul><li>Embed prior knowledge standards within a lesson when they are directly connected to the grade-level standard being taught.</li><li>Use prior knowledge standards to inform future planning and instructional activities.</li><li>Determine whether students have learned the prior knowledge necessary for the current unit's essential standards using a pretest or from working with students in previous units.</li></ul> | <ul><li>Engage students in learning prior knowledge standards before or while teaching grade- or course-level standards in Tier 1.</li><li>Create explicit opportunities to revisit specific prior knowledge content and skills.</li><li>Connect prior knowledge instruction to current learning of essential standards and targets.</li><li>Focus on development of prior knowledge vocabulary and symbols.</li></ul> |

Under No Child Left Behind (NCLB) and Every Student Succeeds Act (ESSA), U.S. states moved to a growth model for state accountability (Klein, 2015). The growth model involves measuring the progress of individual students over time, including a preassessment at the beginning of the year and a postassessment at the end of the year (for example, The Northwest Evaluation Association, MAP® Growth, i-Ready, and so on).

As states moved to a growth model, there was an increase in collaborative teams creating preassessments that mirrored the end-of-unit assessments to measure growth from the beginning of the unit to the end. The challenge with a preassessment that is the same as the postassessment is the preassessment information does not provide quality feedback on current learning. Also, teachers may view this type of preassessment as a waste of time because (1) they are not immediately using evidence of student learning to inform instruction, and (2) students are being assessed on content they have not yet been taught. Instead of learning from the preassessment, collaborative teams teach as originally planned and only use the preassessment data to show growth. Unfortunately, this growth is not very meaningful since students are learning standards for the first time and are expected to show significant growth from the beginning to the end of any unit.

Instead, when needed, use a high-quality preassessment designed to identify students' current understanding of the prerequisite skills needed to learn each essential standard in the unit. This type of preassessment provides information about the appropriate just-in-time supports to include in Tier 1 initial teaching and reveals any targeted interventions needed in Tier 2. Thus, it is a more effective way to use preassessments (Callahan, 2012; Kramer & Schuhl, 2023; Ochsendorf & Pyke, 2007). You and your team will use the information from the preassessment right away to accelerate student learning.

A preassessment that measures prerequisite skills and prior knowledge standards should be short and is only needed when you do not know whether students have learned the prerequisite skills. If you already have this information from prior work with students, a preassessment is not needed.

### Just-in-Time Supports

Schuhl and colleagues (2024) describe *just-in-time supports* as follows: "This means the intervention focuses on essential learning standards students are currently learning or have just learned in the previous mathematics unit" (p. 131). Interventions focus on the immediate prerequisite skills that students need to access the grade-level essential standards being taught at that time.

In their jointly published paper, *Continuing the Journey: Mathematics Learning 2021 and Beyond*, NCTM, NCSM, and ASSM (2021) recommend teachers place less emphasis on dedicating isolated time to reteach previous skills and instead seek opportunities to connect previous learning to current grade-level instruction. For years, and certainly after the pandemic, many teachers began each school year by reviewing or reteaching previous grade-level standards with the intention of filling in instructional gaps that students may have developed. However, it is not necessary to review every part of each standard, and this just-in-case remediation may cause more harm than help (Kramer & Schuhl, 2023; NCTM et al., 2021; New Teacher Project, 2021; Schuhl et al., 2024). Revisiting previous learning without connection to current learning targets may result in students seeing learning opportunities as separated, isolated, and unrelated events, causing them to miss the purpose of the re-engagement (Kramer & Schuhl, 2023).

When students are constructing new knowledge, real-time scaffolding and support helps them see coherence in their learning and, thus, learn grade-level standards without compromising rigor (NCTM, 2014; NCTM et al., 2021; Toncheff et al., 2024). During Tier 2 instruction, you and your team might focus on engaging students in learning any needed prior knowledge in tandem with grade-level content in Tier 1. In chapter 3 (page 43), we discussed that teaching grade-level learning every day also requires a connection to prior knowledge learning. Refer to figure 3.1 (page 47) for examples showing how to teach prior knowledge and grade-level content in Tier 1 and prior knowledge in Tier 2 for a daily learning target.

You and your team may want to address just-in-time supports during Tier 1 instruction. Schuhl and colleagues (2024) offer the following examples for how to address just-in-time supports in Tier 1:

- A lesson or series of lessons the team agrees to implement
- A flex day the team uses to share students, with each team member addressing a particular skill that students need to learn
- A concept knitted into specific and agreed-on lessons for the next unit (for example, continue working on word problems using team agreed-on strategies)
- Beginning-of-class routines focused for two weeks on engaging students in learning the essential learning standard. (p. 128)

Instead of intervening at the start of the year by reteaching random concepts not necessarily connected to the first essential standards being taught, determine as a team how to intervene right away in each unit using frequent common assessment data connected to the standards students are learning (Mattos et al., 2025). You may even use Tier 2 strategies to re-engage students in learning prior knowledge standards before teaching the lessons in Tier 1 requiring that learning.

### Learning Progressions

One question that may come to mind for your team is, What content do we focus on to support our students' prior knowledge? Referring to the quote at the opening of this chapter, teachers need knowledge of the mathematics for their grade or course level, the previous grade or course, and the grade or course that follows (Dixon et al., 2016; SanGiovanni, Katt, Knighten, & Rivera, 2022). Several tools, such as progression guides, curriculum maps, and coherence maps, are available to support teachers in exploring the progression of mathematics content. Digital interactive maps, such as the coherence map on the Achieve the Core website (www.achievethecore.org/page/1118/coherence-map), can help you trace previous- and within-grades standards that support student learning of grade-level content, as well as explore how the standard is connected to future learning.

Figure 4.2 (page 62) shows a learning progression connected to a second-grade standard and an eighth-grade standard. Examine how each grade-level standard (or standards) provides the learning needed for students to progress until they reach their current grade-level standard.

Sometimes, numerous prior grade standards support a single grade-level standard. Teachers then identify the one or two standards most fully connected to that essential standard to make the biggest impact on student learning based on students' needs. The coherence map provides a visual web mapping the progression of mathematics concepts from previous grades' standards to future standards. Some standards in the coherence map on the Achieve the Core website (www.achievethecore.org) also include lessons, tasks, and assessment items for your team to use. Incorporate these resources in instructional moments that focus on supporting students' prior knowledge.

### Prior Knowledge Routines

Beginning your mathematics lesson with a connection to prior knowledge helps students build connections between what they already know and provides a foundation for the learning they will soon encounter (SanGiovanni et al., 2022; Toncheff et al., 2024). Prior knowledge routines establish a context for the day's lesson, which allows students to continue building their understanding and application of the content. It is important to resist simply telling students what they learned in the past that connects to the current lesson; instead, engage them in revisiting prior knowledge learning through tasks, examples, and discussions. Routines "provide a framework for [students'] making sense, reasoning, and discussing" (SanGiovanni et al., 2022, p. 63) and should be brief engagements that become familiar with consistent implementation.

Prior knowledge routines are one of the six elements of high-quality lesson design for teachers to include in daily lessons from *Mathematics Instruction and Tasks in a PLC at Work, Second Edition* (Toncheff et al., 2024). Examples of prior knowledge routines include warm-up tasks or activities, examples, number talks, or a set of problems that support making connections between previous and new learning, to name a few. Figures 4.3–4.6 (pages 64–66) provide examples of prior knowledge routines for kindergarten, grade 1, grade 5, and high school.

## Grade 2 Standard With Prior Knowledge from a Coherence Map

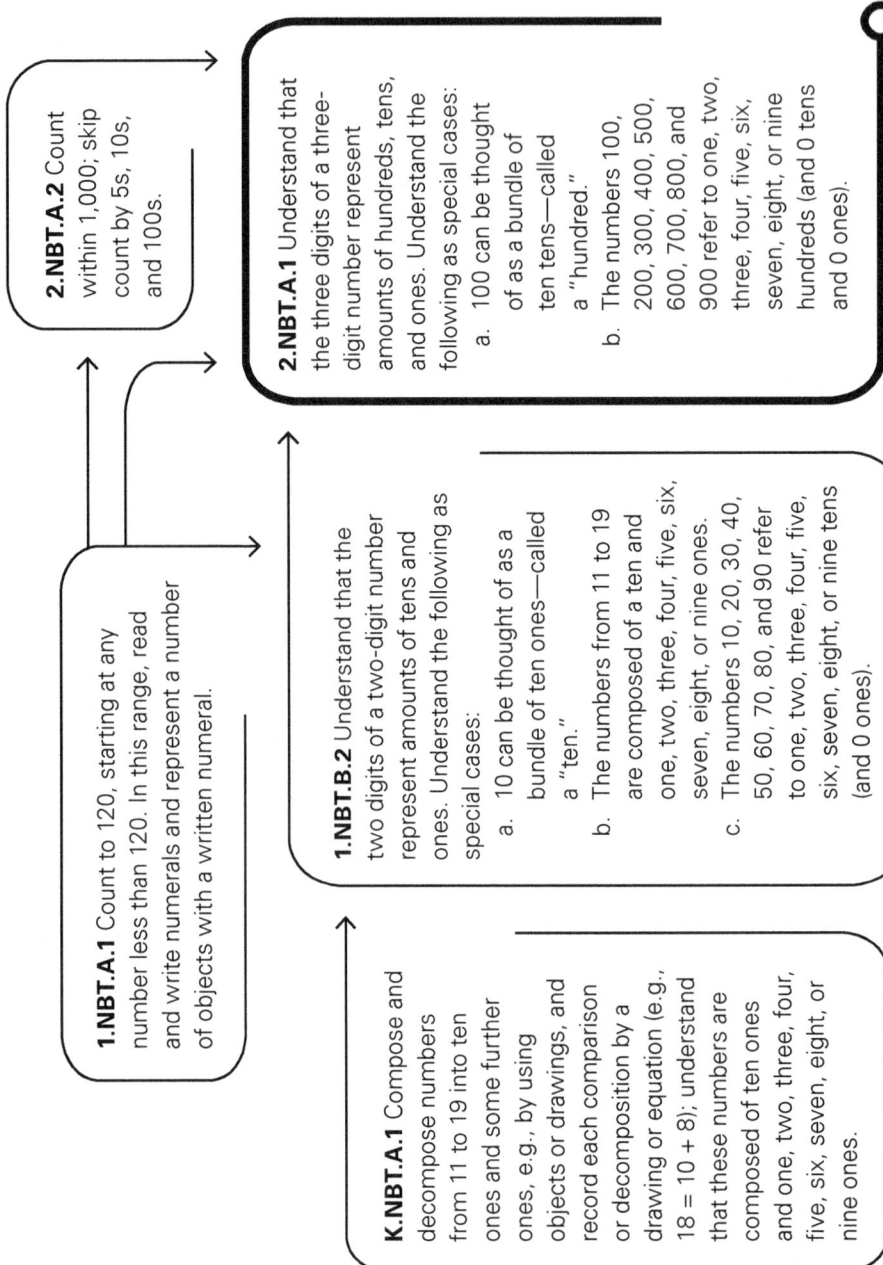

Connect to Prior Knowledge 63

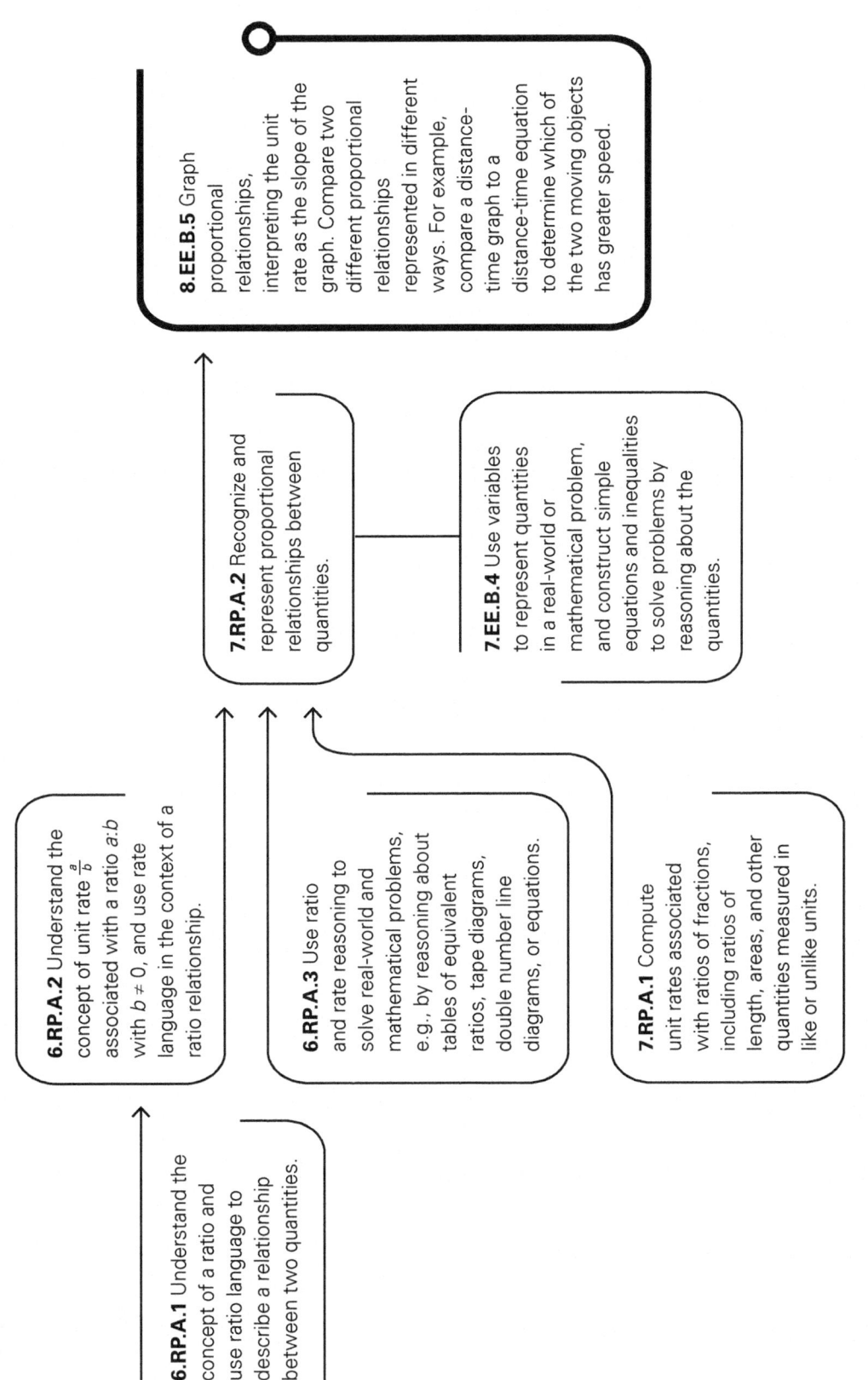

*Source for standards: NGA & CCSSO, 2010.*

**Figure 4.2: Grade 2 and grade 8 standard with prior knowledge in a coherence map.**

| Grade or Course | Grade-Level or Course-Based Standard | Prior Knowledge Standard From Previous Grade, Course, or Unit | Prior Knowledge Sequence Task |
|---|---|---|---|
| Kindergarten | Students should understand the relationship between numbers and quantities and connect counting to cardinality. | **PreK** Students begin to explore one-to-one correspondence situations. | • Pass out toys to 5 friends by giving each child 1 toy and saying the count aloud.<br>• Walk downstairs and count the steps on the way down. |
| | | **PreK** Students recognize numbers in print and small quantities (1–3) by labeling with a number name. | Hunt for specific numbers 1–10 in the environment on signs, toys, clothing, and household or classroom items. |

*Source: Adapted from Toncheff et al., 2024.*

**Figure 4.3: Kindergarten prior knowledge routine example.**

*Visit **go.SolutionTree.com/MathematicsatWork** for a free blank reproducible version of this figure.*

| Grade or Course | Grade-Level or Course-Based Standard | Prior Knowledge Standard From Previous Grade, Course, or Unit | Prior Knowledge Sequence Task |
|---|---|---|---|
| **Grade 1** | 1.OA.C.6 Add and subtract within 20, demonstrating fluency for addition and subtraction within 10. Use strategies such as counting on; making 10 (e.g., 8 + 6 = 8 + 2 + 4 = 10 + 4 = 14); decomposing a number leading to a ten (e.g., 13 – 4 = 13 – 3 – 1 = 10 – 1 = 9); using the relationship between addition and subtraction (e.g., knowing that 8 + 4 = 12, one knows 12 – 8 = 4); and creating equivalent but easier or known sums (e.g., adding 6 + 7 by creating the known equivalent 6 + 6 + 1 = 12 + 1 = 13). | **Kindergarten** K.NBT.A.1 Compose and decompose numbers from 11 to 19 into ten ones and some further ones, e.g., by using objects or drawings, and record each composition or decomposition by a drawing or equation such as 18 = 10 + 8; understand that these numbers are composed of ten ones and one, two, three, four, five, six, seven, eight, or nine ones. | Alex used the following unit cubes to fill a ten frame.<br><br>How many units were left after she finished filling the ten frame? How many unit cubes does she have altogether?<br>Finish the equation:<br>____ + 10 = ____<br>Repeat with different numbers of unit cubes. |
| | | **Kindergarten** K.OA.A.4 For any number from 1 to 9, find the number that makes 10 when added to the given number, e.g., by using objects or drawings and recording the answer with a drawing or equation. | Using a cup, 10 two-color counters, and a ten frame, have students put the counters in the cup, shake it, and pour out the counters. Then ask students to arrange the counters by color in their ten frame and record a drawing or equation to describe the number of red counters and the number of yellow counters that filled the ten frame.<br>For example:<br><br>4 + 6 = 10 |

*Source: Adapted from Toncheff et al., 2024.*
*Source for standards: NGA & CCSSO, 2010.*

**Figure 4.4: Grade 1 prior knowledge routine example.**

Connect to Prior Knowledge 65

| Grade or Course | Grade-Level or Course-Based Standard | Prior Knowledge Standard From Previous Grade, Course, or Unit | Prior Knowledge Sequence Task |
|---|---|---|---|
| Grade 5 | 5.MD.C.5 Relate volume to the operations of multiplication and addition and solve real-world and mathematics problems involving volume.<br>a. Find the volume of a right rectangular prism with whole-number side lengths by packing it with unit cubes, and show that the volume is the same as it would be by multiplying the edge lengths, equivalently by multiplying the height by the area of the base. Represent threefold whole-number products as volumes, e.g., to represent the associative property of multiplication.<br>b. Apply the formulas $V = l \times w \times h$ and $V = B \times h$ for rectangular prisms to find volumes of right rectangular prisms with whole-number edge lengths in the context of solving real-world and mathematical problems.<br>c. Recognize volume as additive. Find the volumes of solid figures composed of two non-overlapping right rectangular prisms by adding the volumes of the non-overlapping parts, applying this technique to solve real-world problems. | **Grade 4**<br>4.MD.A.3 Apply the area and perimeter formulas for rectangles in real-world and mathematical problems. For example, find the width of a rectangular room given the area of the flooring and the length, by viewing the area formula as a multiplication equation with an unknown factor. | Andriel is spreading soil to start a new garden. He wants the garden to be 6 feet wide. What is the longest possible side length for his garden if he has enough soil to cover 84 square feet?<br><br>*? feet*<br><br>84 *square feet*    6 *feet* |
| | | **Grade 3**<br>3.MD.C.7 Relate area to the operations of multiplication and addition.<br>a. Find the area of a rectangle with whole-number side lengths by tiling it, and show that the area is the same as would be found by multiplying the side lengths.<br>b. Multiply side lengths to find areas of rectangles with whole-number side lengths in the context of solving real-world and mathematical problems, and represent whole-number products as rectangular areas in mathematical reasoning.<br>c. Use tiling to show in a concrete case that the area of a rectangle with whole-number side lengths $a$ and $b + c$ is the sum of $a \times b$ and $a \times c$. Use area models to represent the distributive property in mathematical reasoning.<br>d. Recognize area as additive. Find areas of rectilinear figures by decomposing them into non-overlapping rectangles and adding the areas of the non-overlapping parts, applying this technique to solve real-world problems. | Hannah is using square tiles to find the area of a rectangle. Here is the beginning of her work.<br><br>Show at least two different ways to figure out how many tiles Hannah will use to cover the entire rectangle. |

*Source: Adapted from Toncheff et al., 2024.*
*Source for standards: NGA & CCSSO, 2010.*

**Figure 4.5: Grade 5 prior knowledge routine example.**

| Grade or Course | Grade-Level or Course-Based Standard | Prior Knowledge Standard From Previous Grade, Course, or Unit | Prior Knowledge Sequence Task |
|---|---|---|---|
| Geometry | HSG-CO.A.2 Represent transformations in the plane; describe transformations as functions that take points in the plane as inputs and give other points as outputs. Compare transformations that preserve distance and angle to those that do not (e.g., translation versus horizontal stretch). | **Grade 8** 8.G.A.1 Verify experimentally the properties of rotations, reflections, and translations on a coordinate plane. 8.G.A.3 Describe the effect of dilations, translations, rotations, and reflections on two-dimensional figures using coordinates. **Algebra 1** HSF-BF.B.3 Determine the effects on the graph $f(x)$ when $f(x)$ is replaced by $af(x)$, $f(x) + d$, $f(x - c)$, and $f(bx)$ for specific values of $a$, $b$, $c$, and $d$. | Use $\triangle ABC$ in the following graph to answer the questions. 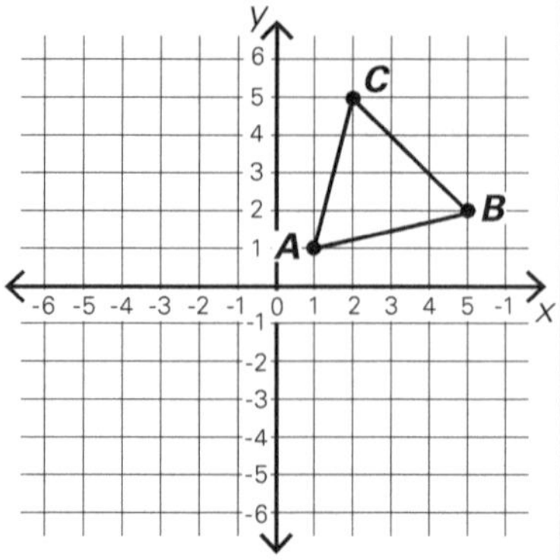 1. Draw a reflection of $\triangle ABC$ over the $y$-axis. Show the ordered pairs for the image $\triangle A'B'C'$. 2. Draw a translation of $\triangle ABC$ down five units and left one unit. Show the ordered pairs for $\triangle A'B'C'$. |

*Source: Adapted from Toncheff et al., 2024.*
*Source for standards: NGA & CCSSO, 2010.*

**Figure 4.6: High school prior knowledge routine example.**

Mona Toncheff and colleagues (2024) offer the following guidelines for designing effective prior knowledge routines.

- Allow students to work in small groups to allow for collaboration and communication opportunities.
- Create student-centered activities from high-level-cognitive-demand tasks, discussions, and prompts that reflect the prior knowledge most connected to the standard and learning targets.
- Structure a clear beginning-of-class routine, allotting no more than five to ten minutes of instructional time; allow slightly more time for more exploratory tasks.
- Assess students through observation and discussions with student groups and avoid using time to review the task. Instead provide examples and responses with explanations for students to review themselves.

For additional prior knowledge routines, see chapter 2 in *Mathematics Instruction and Tasks in a PLC at Work, Second Edition* (Toncheff et al., 2024).

### Centers and Games

If your school has a block of time dedicated to Tier 2 interventions—or your team has planned flex days or teaches on a block schedule—centers and games are engaging ways to reinforce prior knowledge based on students' needs. Games help students develop strategic reasoning and problem-solving skills within essential learning standards and targets. Centers have an element of variability, so students re-engage several times with the same learning structure but with different outcomes (SanGiovanni et al., 2022).

The following are directions for the card game *Salute!* (adapted from Bay-Williams & Kling, 2014). Three players use one deck of playing cards with the face cards removed; aces represent the number one. The purpose of this game is to practice and reinforce multiplication facts and support reasoning using the relationship between multiplication and division. In this game, students draw random cards, identify unknown factors, and explore the relationship between multiplication and division.

1. Players 1 and 2 each select a card face down.
2. Without looking at the card, players place their card on their forehead, facing out to the other player.
3. Player 3 multiplies the two numbers together and calls out the product of the numbers.
4. The first player to correctly identify the number on their forehead wins both cards for that round.
5. Play continues until all the cards have been used or the allotted time has expired.

You can modify this game in various ways, such as finding the sum of the two addends when exploring the relationship between addition and subtraction. After students have had an opportunity to engage in a center or play a game, it is important to have some level of discussion around strategies and reasoning used during play. This allows students to hear their peers' thoughts and continue to refine their own understanding.

Games and centers should be easy for students to understand and engage in; they should provide support and connections to current learning targets. You can even create them from previously used tasks. Planning for *how* students will interact with the game or center and preparing the necessary materials, classroom space, and directions are vital to maximizing students' learning time.

## Conclusion

One important question that both teachers and students should be able to answer is, "How is current learning directly related to previous learning?" It is important to recognize mathematics standards as a coherent progression, with each stop (grade or course) as a previous learning experience needed to support current learning experiences. Tools, such as coherence maps and progression guides, help you and your team identify and plan for student learning of targeted prior knowledge.

Establishing prior knowledge routines in daily lessons helps students strengthen their foundations for new knowledge and helps them build connections between previously learned content and new concepts. As a team, consider preteaching prior knowledge skills in Tier 2 before the unit starts using a pretest on prior knowledge standards. Your team might also re-engage students after a common mid-unit assessment to include a focus on prior knowledge standards leading to grade- or course-level standards.

## Questions to Consider for Next Steps

As a collaborative team, use the following questions to reflect on your current practices and determine any next steps related to focusing on prior knowledge.

1. How will you and your team identify prior knowledge while planning instruction? What tools might you use?
2. How will you and your team determine the specific students who have not yet learned the necessary prior knowledge standards and which prior knowledge standards need to be taught?
3. How will your prior knowledge routines create context for that day's learning target?
4. How can you and your team sustain prior knowledge routines?
5. How might you gather evidence that students are learning prior knowledge through Tier 1 and Tier 2 instruction?

> Bre, a special education teacher for the intermediate grades at an elementary school in Virginia who worked with Jenn, talked to her team about her concerns regarding her students' current understanding of fractions. Some of her observations and analysis of student work included the following.
>
> - Not making the transition from using base ten blocks to model whole numbers to representing decimal values and, therefore, having difficulty with decimal-place-value language and problem solving
> - Recording the numerator and the denominator incorrectly, particularly with fractions greater than one
> - Shading in fraction values but not being able to find equivalent fractions or simplify, even with benchmark denominators, such as halves, fourths, and eighths
> - Struggling to make connections between whole number operations and composing and decomposing fractions
>
> Many educators looking at this list would come to the same conclusion as Bre: These students need extra time and support with foundational whole number and fraction numeracy. Naming students' instructional needs was not Bre's main concern. She was worried about her students being able to access grade-level mathematics in fifth and sixth grade.

CHAPTER 5

# Develop Number Sense

*Children who understand number concepts know that numbers are used to describe quantities and relationships, and are useful tools for getting information about the world they live in.*

—Kathy Richardson

Whether a student is in elementary, middle, or high school, a student's ability to flexibly work with numbers and understand how they relate to one another impacts learning. For the students not yet demonstrating number sense in Bre's classroom (see story on page 68), she now needs to connect foundational number sense of whole numbers to decimals and fractions so students engage in grade-level mathematics. Some foundational number sense can be taught in Tier 1 and earlier number sense in Tier 2.

With number sense, students understand the different ways to regroup numbers, check for the reasonableness of answers, write and represent values using varied equivalent forms (for example, 0.5, $\frac{1}{2}$, and $\frac{6}{12}$), predict answers, and interpret the meaning of ordered pairs on graphs of functions, to name a few. They also, over the course of their preK–12 experiences, connect whole numbers to integers, integers to rational numbers, and then rational numbers to real and complex numbers.

Developing number sense takes time and practice and is addressed daily in Tier 1 instruction, regardless of grade. It becomes a focus for Tier 2 instruction when learning moves forward, but some students are not yet demonstrating the grade-level number sense required from prior units. An understanding of whole numbers (for example, less than, greater than, compose and decompose numbers, and values on a number line), place value to 100, and addition and subtraction with and without regrouping may become Tier 3 learning experiences after second grade (NCTM, 2000).

## The *Why* and the *What* of Developing Number Sense

Building number sense is foundational to all preK–12 mathematics learning. From an early understanding of numbers developed through counting—as well as composing and decomposing numbers—to navigating real and complex numbers, students who flexibly work with numbers develop number sense (NCTM, 2000, 2020). In *Principles and Standards for School Mathematics*, NCTM (2000) states:

> As they progress from prekindergarten through grade 12, students should attain a rich understanding of numbers—what they are; how they are represented with objects, numerals, or on number lines; how they are related to one another; how numbers are embedded in systems that have structures and properties; and how to use numbers and operations to solve problems. (p. 32)

British researchers Eddie M. Gray and David O. Tall (1994) asked teachers to identify students who they considered low, middle, or high achieving. After students demonstrated skills with operations of numbers, the researchers concluded those designated as low-achieving students did not know less about mathematics, rather they were interacting with mathematics in a more difficult way because they were not demonstrating flexibility with numbers (Boaler, 2016). Students focused on algorithms when it did not make sense to do so. For example, instead of finding 23 − 5 by breaking 5 into 3 and 2 to reach 18, students used an algorithm, which was more difficult. Jo Boaler (2016), mathematics professor at Stanford and co-founder of YouCubed, also finds in her work that students considered low are not making sense of mathematics or developing number sense, which later impedes their learning progress.

Number sense goes well beyond an understanding of the base ten system; it includes developing a sense of magnitude, thinking flexibly to compose and decompose values, and extending understandings of whole numbers to fractions, decimals, integers, rational numbers, real numbers, and complex numbers (Boaler 2016; Burns, 2007; NCTM 2000, 2020). Students with number sense are intuitive when solving problems, can estimate, and are better at determining the reasonableness of answers (Van de Walle, Karp, & Bay-Williams, 2019).

When students have not developed number sense, they tend to struggle to understand grade-level skills and concepts. Mathematics educator Marilyn Burns (2007) shares, "Students with poor number sense tend to rely on procedures rather than reason, often do not notice when answers or estimates are unreasonable and have limited numerical common sense" (p. 24). It is crucial for students to develop number sense starting in preK and continue growing that understanding through high school.

There are important benchmarks to reach when developing numbers sense, as shown in table 5.1. Within each grade band, number sense is taught as Tier 1 and possibly as Tier 2 content, if students have not learned it in the designated grade or course. Later, if students are still working to develop number sense from several grades prior, in addition to Tier 1 and Tier 2 experiences for grade- or course-level standards (and possibly prior knowledge standards), use Tier 3 experiences to build number sense. See chapter 1 (page 9) for the connection between the three tiers of RTI and the team ownership of Tiers 1 and 2.

By understanding students' numeracy strengths and needs, you can develop plans to bolster number sense through targeted whole-group and small-group instruction. Table 5.2 shows some keys to building number sense in Tier 1 core instruction and Tier 2 interventions.

You and your team determine which skills to teach and develop during Tier 1 core instruction, which impacts how students learn mathematics concepts, such as number sense and procedural fluency. Often, students develop number sense through the strategies used to teach operations in class, but you still need to intentionally teach and address it in lessons. For the purposes of this book, we focus on Tiers 1 and 2.

Table 5.1: Number Sense by Grade Band

| Grades PreK–2 | <ul><li>Count objects to tell how many.</li><li>Use models to develop place value understanding.</li><li>Compose and decompose whole numbers.</li><li>Connect number names with numerals.</li></ul> |
|---|---|
| Grades 3–5 | <ul><li>Represent and compare whole numbers and decimals.</li><li>Recognize and generate equivalent representations for whole numbers and fractions by composing and decomposing numbers.</li><li>Understand the meaning of a fraction and its magnitude.</li></ul> |
| Grades 6–8 | <ul><li>Work flexibly with fractions, decimals, and percents.</li><li>Compare and order fractions, decimals, and percents.</li><li>Understand and use ratios and proportions to represent relationships.</li><li>Understand and work with large numbers using scientific notation.</li></ul> |
| Grades 9–12 | <ul><li>Understand very large and very small numbers.</li><li>Compare and contrast number systems (whole numbers, integers, rational, real, and complex).</li><li>Understand real and complex solutions.</li></ul> |

*Source: Adapted from NCTM, 2000.*

*Visit **go.SolutionTree.com/MathematicsatWork** for a free reproducible version of this table.*

Table 5.2: Develop Number Sense in Tier 1 and Tier 2

| Develop Number Sense in Tier 1 and Tier 2 |
| --- |
| • Connect counting to place value and represent whole numbers, fractions, decimals, and integers.<br>• Use models when developing early understanding of numbers and intentionally connect concrete models to more abstract representations of numbers and values.<br>• Engage in subitizing routines to see and support the unitizing of values.<br>• Build an understanding of the four operations, and the relationships between the operations, through contexts and exposure to varied strategies rather than following rote procedures.<br>• Assess the reasonableness of answers from a word or mathematical problem by explaining whether an answer makes sense during whole-group and small-group student discourse. |

| Develop Number Sense in Tier 1 | Develop Number Sense in Tier 2 |
| --- | --- |
| • Offer students access to choosing strategies, tools, and models to work with when developing numeracy.<br>• Use multiple strategies to compose and decompose numbers and when learning to compute with the four operations.<br>• Compare and order numbers using a number line.<br>• Connect place value concepts to the four operations.<br>• Show multiple equivalent ways of writing a number (for example, 1.5, $\frac{3}{2}$, and $\frac{12}{8}$).<br>• Predict the answer to a mathematical task. | • Ensure that working with concrete models during re-engagement coincides with instruction focused on writing mathematical thinking using numbers and symbols.<br>• Connect subitizing routines to the use of mental mathematics strategies and how to compose and decompose numbers.<br>• Support retention of numeracy skills by incorporating independent practice and partner work through games.<br>• Repeatedly reference the magnitude of an answer using a number line.<br>• Focus on generating equivalent representations for numbers (for example, 31 is 3 tens and 1 one, or 2 tens and 11 ones, or 1 ten and 21 ones, or 31 ones). |

## The *How* of Developing Number Sense

There are many ways to build number sense during Tier 1 and Tier 2 instruction. To develop number sense, spiral many of the strategies shared in this chapter through lessons and use them repeatedly with different numbers, as well as in mass practice.

### Early Numeracy

Students are expected to identify a numeral, name a quantity, represent a numerical amount, and compare small values as early as preK and kindergarten. Students need repeated exposure and explicit instruction that often includes simultaneous use of the concrete, symbolic, and oral language that supports early numeracy and counting. The following strategies show examples of how to develop a beginning understanding of numbers.

*Name and Number Sort*

For this intervention strategy, create sets of cards with the numbers 0 to 10 and the letters in each student's first name. Use a template, such as the one in figure 5.1, to prepare a set of cards for the entire class or for a small group with this specific need.

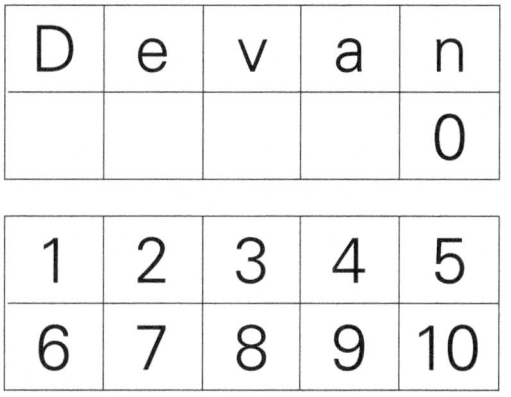

**Figure 5.1: Example name and number sort.**

After modeling and guided practice, students use their set of cards to name and order both the numerals through 10 and the correct spelling of their name.

### Give Me a Quantity

Students in the primary grades often engage in counting routines involving the rote forward sequence, like counting from 1 to 10 and counting objects in organized ways. It is important that you and your team discuss avenues for supporting students who are struggling to master these earliest skills. The Give Me a Quantity strategy not only emphasizes ways to support students counting using one-to-one correspondence, but it also builds an understanding of cardinality and conservation of number. Figure 5.2 shows an example of how you might intentionally teach one-to-one correspondence, cardinality, and conservation of number.

One-to-one correspondence and cardinality teach students that numbers can represent a quantity of objects, which gives meaning to a number's size. Later, numbers may represent quantities, such as measurements, rates, probabilities, or function values. When students demonstrate conservation of number, they know the number of objects counted remains the same when the objects counted are shuffled and, thus, do not need to be recounted. Understanding one-to-one correspondence, cardinality, and conservation of number leads to students subitizing, or quickly seeing a quantity of objects or dots, and developing structures for knowing how many are shown without needing to count them.

## Subitizing Routines

Subitizing structured number images and patterns supports early fluency, building a system within ten and the concept of multiple groups. Students develop an understanding of number sense as they listen to, make sense of, and use different strategies to quickly identify the number of objects they see. Using five, ten, twenty, and one-hundred frames helps students understand anchor numbers, which leads to a foundational understanding of the base ten number system. Later, teachers use subitizing to support an understanding of fractions and decimals, and in high school as a way to connect key features of graphs to their functions.

### See, Say, Write

All students in the early elementary grades need consistent exposure to dot images (such as those found on dice and dominoes), but students presenting gaps

| Give Me a Quantity | | |
|---|---|---|
| **One-to-One Correspondence** | **Cardinality** | **Conservation of Number** |
| Ask the student to count out a certain number of cubes, bears, or other objects. For example, state, "Give me eight!"<br><br>**Specific Scaffolds:**<br>Have students count cotton balls using plastic grippers, like tweezers, to pick up and count to slow down the action and focus on one at a time. Have students drop the cotton balls into a ten frame made from an egg carton. | Ask the student to say the number just counted. "How many did we just count?"<br><br>**Specific Scaffolds:**<br>Use the following sentence frame as part of your routine.<br><br>I just counted _____. | Rearrange the number of objects without taking any away and ask, "How many do we have?"<br><br>**Specific Scaffolds:**<br>Use a ten frame to reinforce seeing the number counted and the number missing. 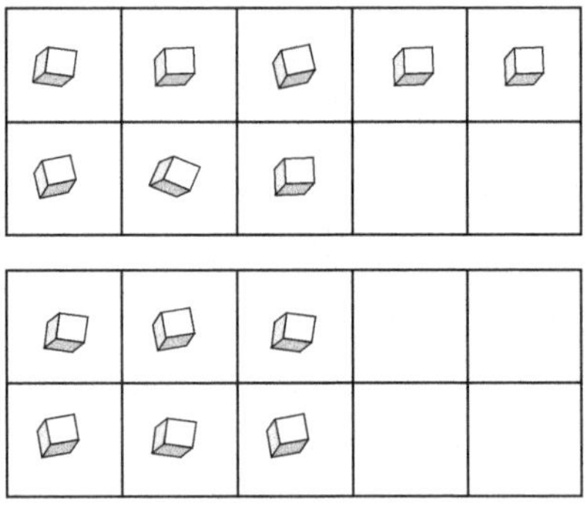 |

Figure 5.2: Give Me a Quantity example.

*Visit* **go.SolutionTree.com/MathematicsatWork** *for a free blank reproducible version of this figure.*

in numeracy have a critical need for structured dot images and the ten and twenty frames. Intervene with structured dot images to reinforce subitizing during whole-group or small-group prior knowledge warm-ups. However, also use structured dot images to engage students in independent practice or cooperative learning. This QR code provides access to materials and activities for the See, Say, Write strategy using subitizing.

Teachers use subitizing in the primary grades to help students quickly determine structures to identify quantities without having to count each individual object. Starting in third grade, subitizing is used to develop a quick way to see whole numbers, as well as fractions. Instructional coach and author Steve Wyborney (n.d.) has developed a website (www.stevewyborney.com) that contains slides to practice subitizing in the primary and intermediate grades.

Additionally, Wyborney (n.d.) created an activity called Splat! used with whole numbers and fractions to develop number sense through composing and decomposing numbers and quick images. In Splat!, teachers show students a slide with the total number of dots or fractional amounts and a slide with a few dots or fractional amounts covered by a black "splat." Students use the images and total value to determine what is under the splat. Figure 5.3 shows examples of Splat! images.

Use images of base ten blocks to show decimal values or percentages and have students quickly identify the value shown. In high school, images of functions also contribute to an understanding of numbers and the numbers used to write functions from graphs. Specifically, you might show pictures of polynomial functions and ask students to determine the most likely power of the function based on the number of zeros shown and the end behavior of the graph. Figure 5.4 (page 74) shows examples using base ten blocks and graphs to build number sense by subitizing.

Quick images used for subitizing help students develop number sense through an understanding of quantity in structured images, whether the image is shown in part or in full. Students visualize and begin to compose and decompose numbers based on how they mentally group the dots or models.

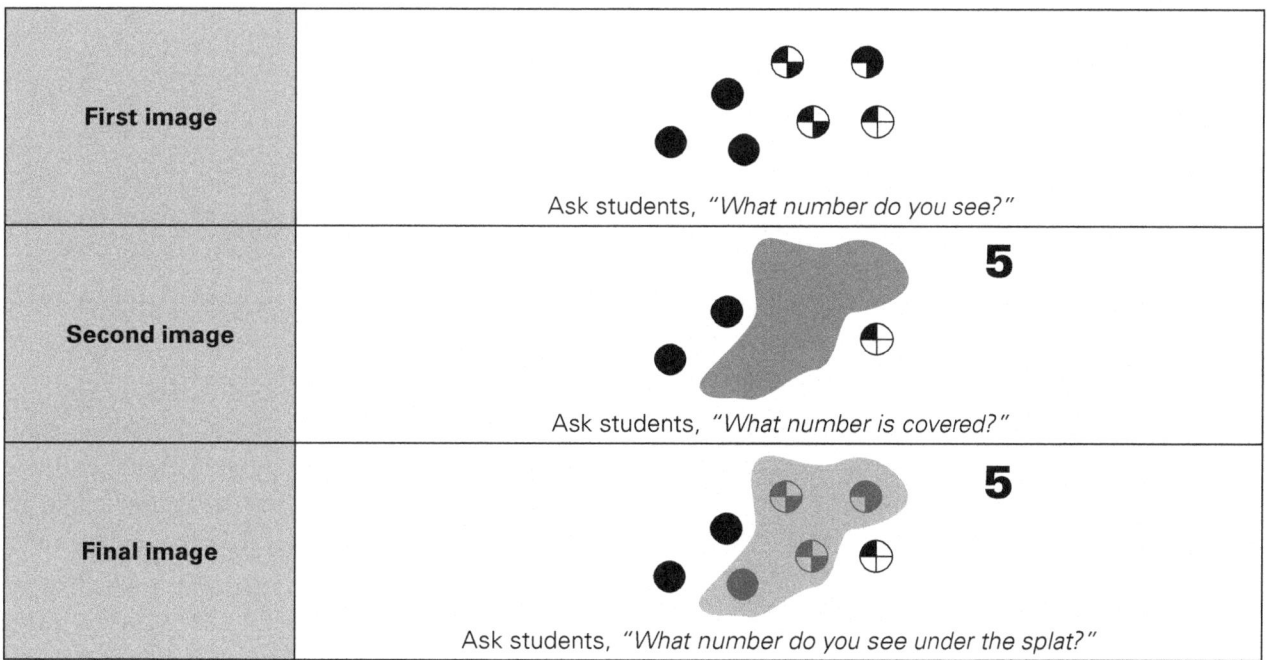

*Source: Adapted from Wyborney, n.d.*

**Figure 5.3: Example Splat! images.**

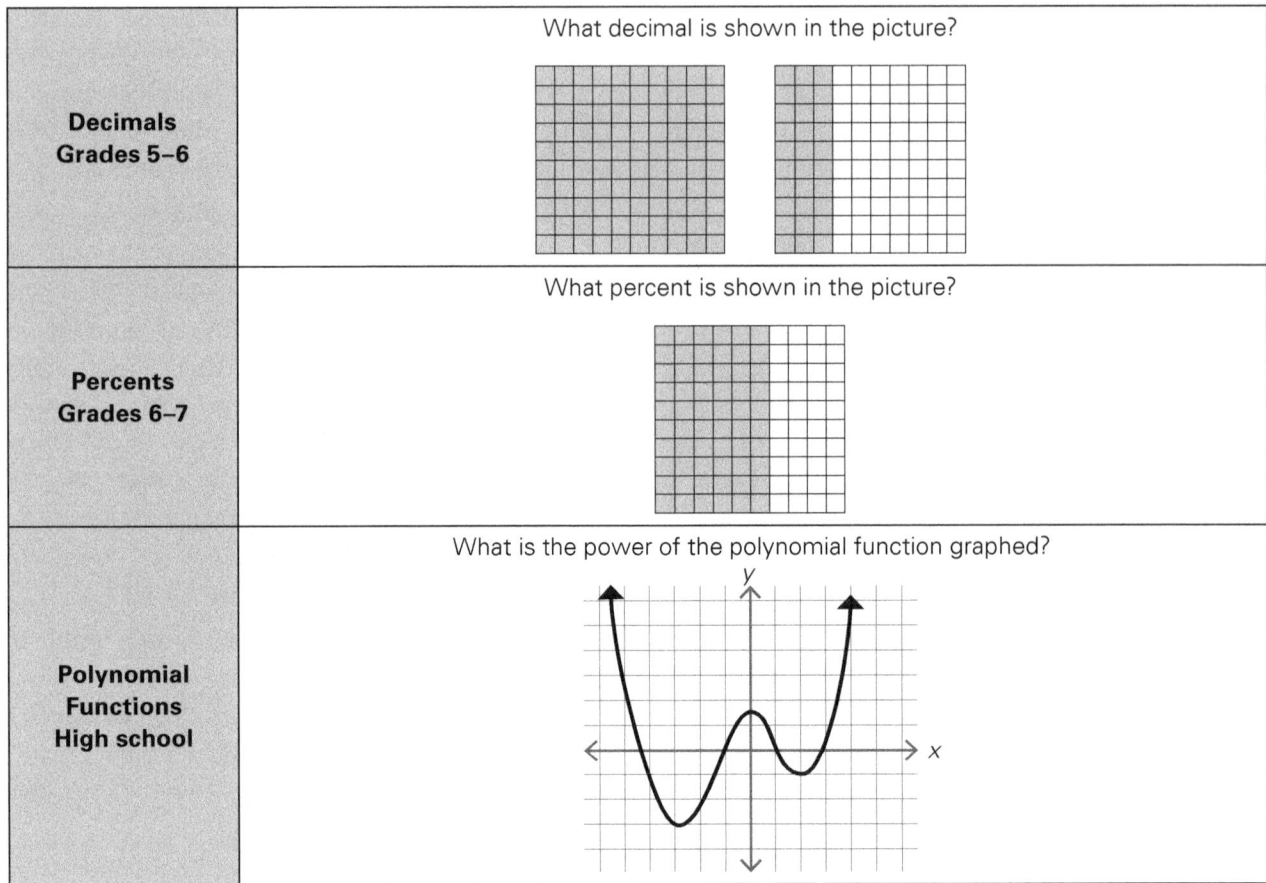

**Figure 5.4:** See, Say, Write examples for grades 5 through high school.

*How Full? How Empty?*

As students begin to make sense of fraction and decimal models, they also need to subitize, or instantly see a quantity, and use their sense of magnitude (size and scale) to name parts of a whole. The How Full? How Empty? activity asks students to evaluate a real-life model and estimate the fractional amount shown or the fractional amount missing. Students might be asked to give their answer as a fraction, decimal, or percent.

Figure 5.5 shows examples of images to use for the How Full? How Empty? strategy. Students connect the image to their knowledge of decimals and percentages using base ten tools or fractions using fraction bars. They might also use other tools to estimate and justify their thinking in concrete ways.

The How Full? How Empty? strategy stimulates student-to-student discourse as students compare their estimates with one another (see chapter 9, page 129, for more information about student discourse). For more meaningful discourse, elementary learners need sentence frames, such as the following.

- The fraction I see is _____.
- The object is _____ full.
- The object is _____ empty.

Using specific sentence frames, elementary learners engage in discussions to learn from one another.

### Forward and Backward Counting Strategies

Fluent use of the forward and backward counting sequences is a critical numeracy skill that continues throughout the elementary grades. Authors Robert J. Wright, Garry Stanger, Ann K. Stafford, and Jim Martland (2015) emphasize, "Many children in the middle and upper primary years make extensive use of counting when solving arithmetical problems, that is, problems that involve any of the four operations" (pp. 52–53). Counting forward and backward helps students understand numbers on a number line which, in turn, leads to students showing addition, subtraction, multiplication, and division on a number line. Additionally, fluid counting combined with

**Figure 5.5: How Full? How Empty? strategy example tasks.**

understanding the concept of before and after impacts a student's ability to determine numbers in between and estimate sums, differences, products, and quotients.

Start with different numbers when counting forward and backward. For example, ask students to count backward from twenty or count forward from thirty-four. Using strategies to develop counting forward and backward fluently is critical to growing students' number sense and their eventual ability to add and subtract (Common Core Standards Writing Team, 2011).

### Counting Off the Decade

Counting by ones to find a sum or difference is reasonable for smaller numbers but cannot efficiently be applied for counting or computing multidigit numbers. While many students use rote counting patterns to count by fives, tens, or even hundreds, some need support counting forward and backward by ten, especially when transitioning to numbers over one hundred. The Counting Off the Decade strategy asks students to count forward or backward by ten.

Students can represent numbers concretely using base ten blocks or beaded number lines (for example, a rekenrek) and make connections to the patterns found on a number chart to make sense of whole-number counting sequences, as shown in figure 5.6. The tools support students' understanding of what number comes next when their knowledge of the pattern breaks down.

Using laundry clips and note cards, students keep track of the count, notice patterns, and transition from counting by ones to counting by tens. Utilizing a 120 or 200 chart, along with base ten models, allows learners to see the pattern beyond 100.

### Counting Fractions With Pattern Blocks

While developing an understanding of the part-to-whole relationship of fractions, give students numerous experiences forming and breaking apart one whole into equal parts. Use mathematics manipulatives, such as pattern blocks or fraction bars. By first naming which tool is one whole, students then use that whole to build values greater than one, counting as they add on blocks. Counting back to zero not only reinforces the counting sequence, but also builds prerequisite knowledge to support subtracting fractions. Facilitate conversations about equivalent fractions or decimals by asking students, "Can we name this value a different way?" An example of this counting strategy is shown in figure 5.7 (page 76).

You may notice students count on from 1 saying $1\frac{1}{4}$, $1\frac{1}{2}$, $1\frac{3}{4}$, 2, $2\frac{1}{4}$ or $\frac{5}{4}$, $\frac{6}{4}$, $\frac{7}{4}$, $\frac{8}{4}$, $\frac{9}{4}$, among other options. There are different ways to say and write equivalent fractions, and it is insightful to see how students are making sense of fractions through their counting.

**Figure 5.6: Counting Off the Decade strategy on a beaded number line.**

**Directions:**
1. Give each student in a small group two to three of the same-sized pattern block or counting rod.
2. Give each group of students a mat to keep track of the count and see the size of the whole.
3. Students take turns adding a piece to the mat, saying the count, and naming equivalent values along the way.
4. Record the count on a number line as students are counting.

| Guiding Questions |
|---|
| Can anyone name this value a different way? |
| What is our total value now? |
| **Sentence Frames** |
| _____ is the same as _____. |
| _____ is equal to _____. |

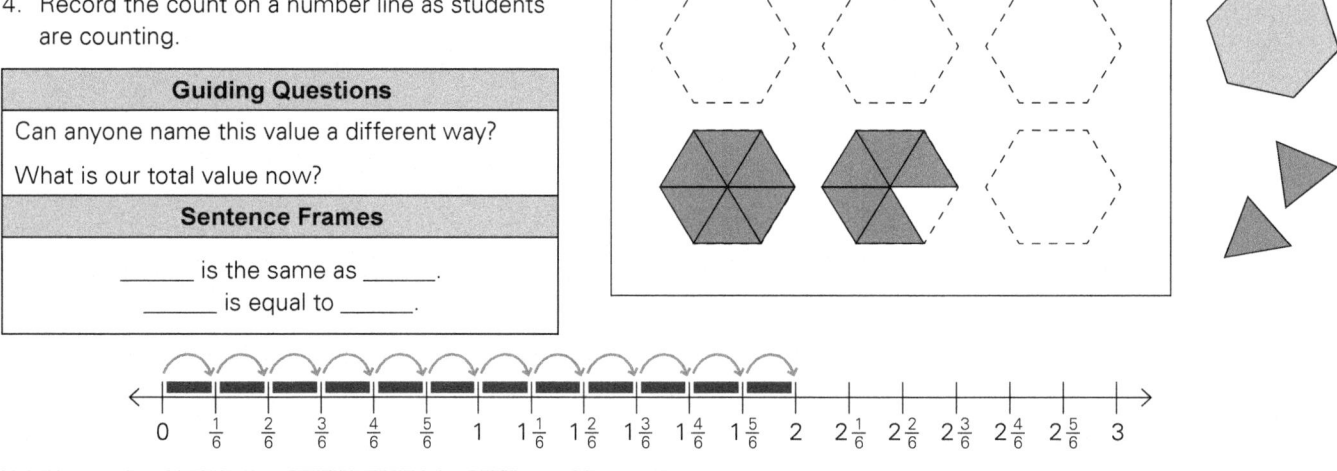

**Figure 5.7:** Counting fractions routine with pattern blocks.

### Number Lines

A number line is a tool that students use to model and represent numbers in ways that compare values, estimate and reason about the size of a number, describe patterns, and more. Use number lines with primary learners to develop an understanding of whole-number sequences, with intermediate students to develop an understanding of the relationship between fractions and decimals, or in secondary mathematics courses to make sense of operating with real numbers and show solution sets to inequalities, to name a few. Additionally, two number lines drawn perpendicular to one another create the coordinate plane used in secondary grades and courses.

In preK and kindergarten, number lines might begin as number tracks, also called number paths, while students are still learning to count and develop a sense of one-to-one correspondence and cardinality (Norris & Schuhl, 2016). Later, in kindergarten and first grade, number lines show whole numbers up to twenty, and every tick mark represents a whole number.

Beginning in second grade, students might use open number lines, which do not include every tick mark uniformly spaced, to show their thinking when adding or subtracting with larger numbers. In third grade, fractions are shown on a number line and scales are used to space tick marks. Decimals are shown in fourth and fifth grade. In sixth grade, negative integers, fractions, and decimals are shown on a number line, and the full coordinate plane is introduced.

Beginning in sixth grade, the number line might be a model that shows all the solutions to an inequality or a single solution to an equation. In high school, the number line is used for all real numbers, including making sense of square root values and estimating between which two whole numbers the value lies. Figure 5.8 shows examples of number lines used in preK–12.

The following are examples showing how your team might engage students in using a number line to show a specific number and its relationship to other numbers.

- Which two tens is the number 72 between? Use the beaded number line to convince your partner.
- Extend the pattern 498, 493, 488, 483, and show your thinking on a number line. How can you describe this pattern?
- Show the following values using a number line to order them from least to greatest: $\frac{2}{3}$, 0.6, 62 percent.

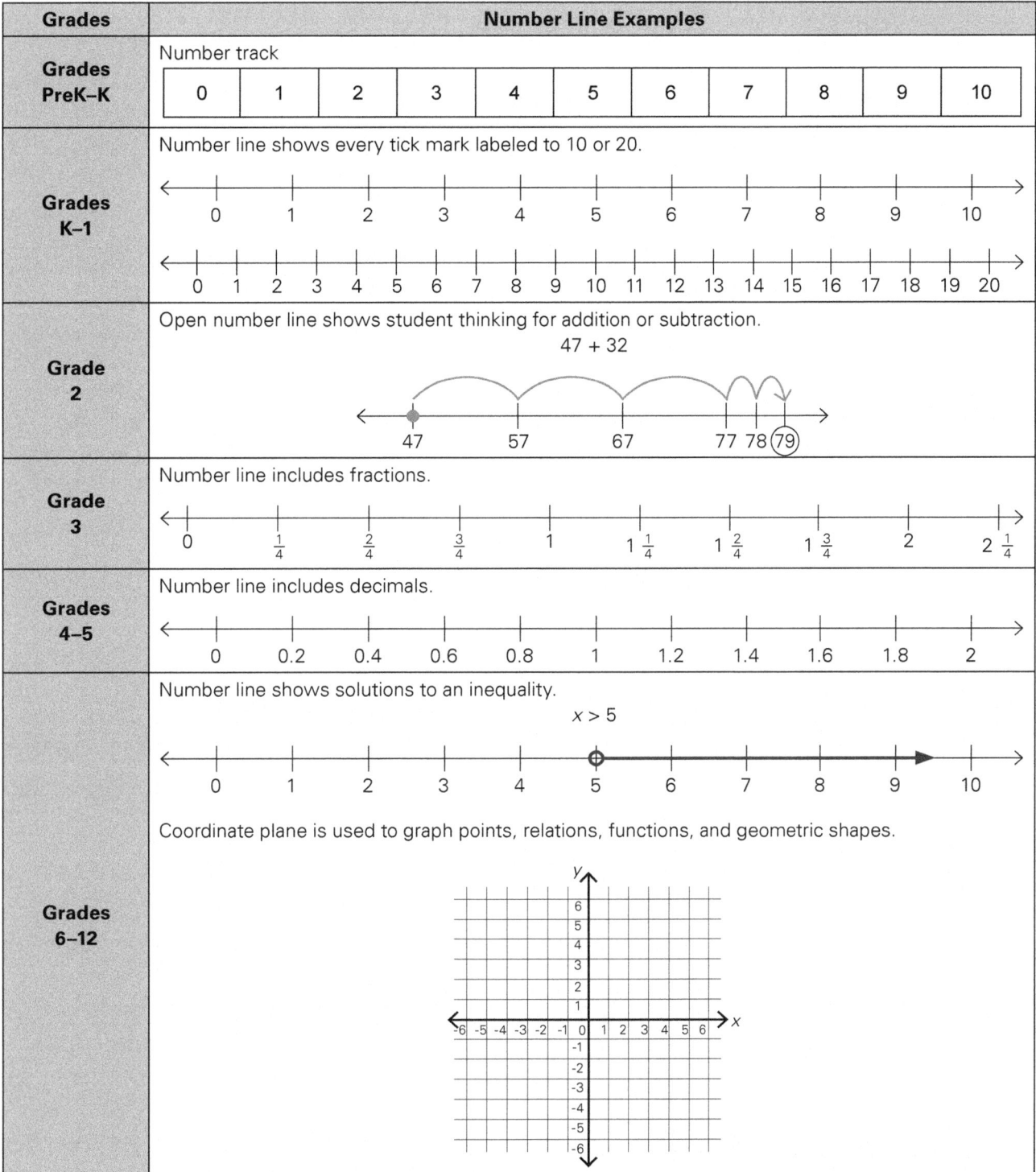

Figure 5.8: Examples of number lines across grade levels.

*Visit **go.SolutionTree.com/MathematicsatWork** for a free blank reproducible version of this figure.*

- Edwin said $\frac{99}{100}$ is the closest number to one whole. Do you agree or disagree? Use a number line to defend your claim.
- Show |5| on a number line.

- Marissa's goal is to save at least $300 this year from her babysitting money but is hoping to save more to go on vacation with her friends. Use both algebraic symbols and statements to express this context and represent the expression on a number line.

- A parking garage charges $10 for the first hour and $4 for each hour after. Graph the line showing the rate in terms of hours. What does the point (6, 30) on the graph represent in terms of the money spent at the parking garage?
- Estimate the location of the following numbers on a number line: $\sqrt{2}$, $\frac{\sqrt{3}}{2}$, $\frac{\sqrt{2}}{2}$, and $\sqrt{17}$.

Initially, in elementary mathematics, representing numbers on a number line during Tier 1 instruction is paired with additional tools, such as base ten blocks or pattern blocks. In middle school and high school, students apply understandings of abstract concepts, such as inequalities, by representing expressions that show an infinite number of values as the solution on a number line. Students in high school will also show limits of functions on a coordinate plane.

During Tier 2 instruction, you may need a stronger emphasis on connecting concrete models to prelabeled number lines with explicit connections to the open number line. Virtual tools, such as those found at https://apps.mathlearningcenter.org/number-line and www.mathsisfun.com/numbers/number-line-zoom.html, allow teachers and students to represent a variety of values on a number line with customizable features. Consider posting a number line made of yarn on the wall. Periodically, have students use small sticky notes to show the placement of their answers to a task on the number line to better make sense of their answers and see equivalent values written in different ways (for example, 2.25 and $2\frac{1}{4}$).

### Number Talks

An ideal strategy for routinely engaging students in building foundational numeracy is number talks, created by educators Ruth Parker and Kathy Richardson in the 1990s (Mathematics Education Cooperative, 2024) and developed by Cathy Humphreys and Sherry Parrish (2014). Number talks are designed to create daily discussions around numbers and operations in a classroom environment. Students share their thinking that led to an answer, and the teacher records their ideas. Students not yet able to perform a computation or make sense of a number then choose an idea to try with a similar follow-up question. Number talk discussions encourage risk-taking behavior and sharing, allowing students to make sense of mental strategies and build bridges toward equitable access to mathematics and, ultimately, develop fluency.

Parrish (2010) explains that "the problems in a number talk are designed to elicit specific strategies that focus on number relationships and number theory" (p. 5). Figure 5.9 shows grade-level examples of number talks.

Number talks are additional practice designed to connect to students' prior knowledge (see chapter 4, page 57, for more information about prior knowledge). We recommend number talks take less than ten minutes during a lesson, and, when implementing them, you are responsible for facilitating the sharing of student thinking, strategies, and reasoning. Intentionally planning for number talks by setting time aside and ensuring structures are in place to build number sense allows you to reinforce desired concepts.

### Estimation and Reasonable Answers

Mathematicians tend to predict what they think the solution to a problem should be before solving it. They also look at their answer to a task and ask themselves if the answer makes sense based on the context or numbers involved. Asking students to predict or estimate an answer and determine whether the answer is reasonable also supports building number sense.

For example, in fourth grade you might give students the following task.

> Jayson is planning a fourth-grade field trip to the science museum. There are 28 students in Mr. Smith's class, 30 students in Mrs. Bates's class, and 25 students in Ms. Farmer's class. The school has small buses to use for field trips. Each bus can hold 15 students. How many buses does Jayson need to schedule for the field trip?

Once students have read and made sense of the task, ask, "Do you and your partner think Jayson will need more or fewer than 7 buses? Why?" or "How many buses do you think Jayson will need?" If asking the latter question, put answers on the whiteboard, and then have students compare their answer to the answers predicted after they have solved the task.

Students can also estimate a solution to a problem involving only numbers. For example, when students are learning to add and subtract in kindergarten, you

| Grade | Question or Prompt |
|---|---|
| Grades PreK–K | How many dots do you see?<br>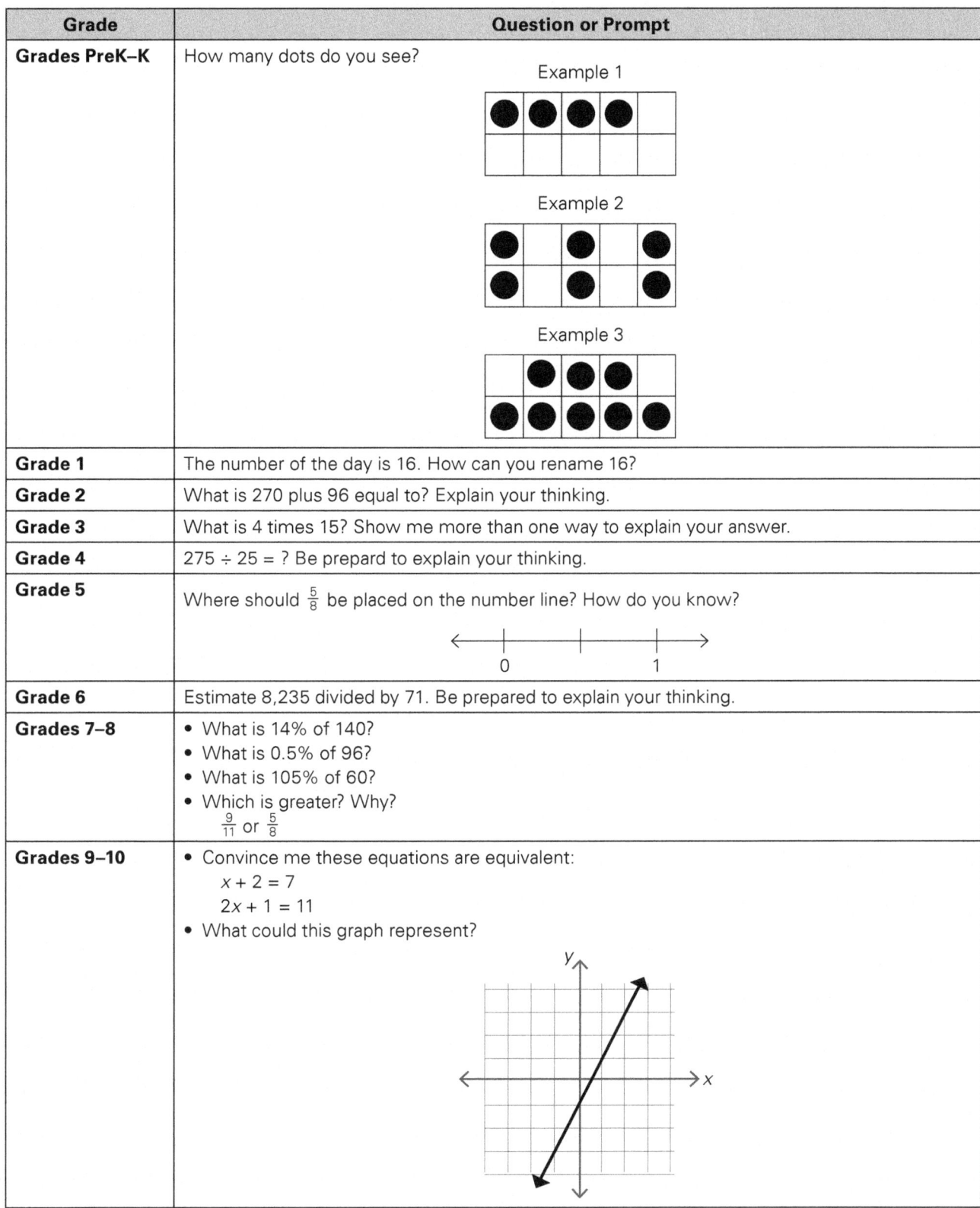 |
| Grade 1 | The number of the day is 16. How can you rename 16? |
| Grade 2 | What is 270 plus 96 equal to? Explain your thinking. |
| Grade 3 | What is 4 times 15? Show me more than one way to explain your answer. |
| Grade 4 | 275 ÷ 25 = ? Be prepared to explain your thinking. |
| Grade 5 | Where should $\frac{5}{8}$ be placed on the number line? How do you know? |
| Grade 6 | Estimate 8,235 divided by 71. Be prepared to explain your thinking. |
| Grades 7–8 | • What is 14% of 140?<br>• What is 0.5% of 96?<br>• What is 105% of 60?<br>• Which is greater? Why?<br>  $\frac{9}{11}$ or $\frac{5}{8}$ |
| Grades 9–10 | • Convince me these equations are equivalent:<br>  $x + 2 = 7$<br>  $2x + 1 = 11$<br>• What could this graph represent? |

**Figure 5.9: Number talk prompt examples.**

might ask them if 3 + 2 is greater than 3 or less than 3. In high school, you might ask students to determine between which two whole numbers is $\sqrt{6}$.

It is good practice to have students stop and ask themselves if their answers make sense. When solving $3 + x = 25$, if a student's answer is 10 or 40, does that make sense? Have students look at their answer in the context of the problem to reason if it is accurate or not.

Students sometimes input an incorrect number or operation when using a calculator and get an answer that does not make sense, but they insist it is correct because the number is on the calculator. Asking for reasonableness will help students identify if the answer makes sense.

Too often, students complete a problem and move on, not stopping to see if their answer makes sense. They also seldom ask themselves at the start of the problem what answer might make sense. Teach students to estimate and check for the reasonableness of their answers as a way to develop number sense.

## Conclusion

The strategies shared in this chapter are designed to build number sense with an emphasis on engaging students in learning experiences that provide opportunities to simultaneously use tools, pictorial models, and verbal and written expressions. Giving primary learners ample opportunities to engage in number routines, such as subitizing and counting, supports building the number sense foundation necessary for learning future mathematics standards in all strands or domains in each grade or course level.

As students age, some may need Tier 1 instruction on grade-level standards, Tier 2 instruction as intervention for grade-level standards, and Tier 3 supports for early numeracy and number sense skills. When students in a middle school or high school course are not yet demonstrating early number sense, consider how to spiral in early-grade number talks and strategies at the start of each lesson (prior knowledge routine for skills students are still learning from years prior). Also, consider how to use small groups or stations to teach early number-sense skills, while students are still learning grade- or course-level standards and getting additional supports in Tier 3.

Engaging learners in Tier 1 involves using strategies to develop number sense while introducing grade- or course-level concepts during best first instruction. Students engaged in Tier 2 learning may need extended guidance, time, and practice to apply number sense strategies to tasks while building strong foundations in numeracy. Strong number sense develops the flexibility needed to access grade- and course-level mathematics.

## Questions to Consider for Next Steps

As a collaborative team, use the following questions to reflect on your current practices and determine any next steps for building number sense.

1. What strategies do you currently use in your classroom or within your teacher team that engage learners in building number sense?

2. What strategies, tools, or models can you and your team use to support numeracy?

3. How will you ensure students are making sense of numbers rather than mimicking rote patterns?

4. What big mathematics concepts are students learning, and how can your team support the concept with a numeracy skill?

5. How are students encouraged to think flexibly when learning mathematics?

> One member of an elementary mathematics team Dr. Buck was coaching, Mr. Davis, wanted Brian's opinion and insights on how he could gain more student success with problem solving. Mr. Davis explained that he implements the same or similar routine every week. The beginning of each week starts with teaching students the week's new skill by showing and explaining how to solve the problems. He then has students complete example problems with him. Occasionally, he uses a manipulative or drawing to show each step, and every so often, he allows students to play with the manipulatives to have a little fun. By the middle of the week, students are beginning to practice the skill independently, still heavily led by Mr. Davis. Occasionally, students' practice includes a couple of real-world problems at the end of the problem set.
>
> Mr. Davis shared he had been teaching students to use a word problem decoding process that encourages them to look for key words and important numbers in a word problem to solve it. When he sees students struggling with word problems, he walks them through using that decoding process. He assigns homework almost every weeknight, except on Fridays. On Fridays, the students take their weekly quiz or test on the week's content.
>
> Mr. Davis observed that when he grades the weekly assessment, a few students usually surpass expectations, many of his students perform at an average level, and the remaining students consistently perform below expectations. He also noticed that the word problems requiring a constructed response are the most troublesome for students. Too many of them provide a response that is not reasonable for the problem or do not provide a response at all.
>
> Mr. Davis explained that when he grades assessments, he leaves feedback by circling and underlining the key words and important numbers that students should have addressed. Although students practice their problem-solving process with him, and he shows them the words and numbers to use, they are still not successful with problems placed in a context that requires them to apply the skills they are learning. "What else can I do?" Mr. Davis asked.

# CHAPTER 6

# Focus on Problem Solving

Problem solving is founded in curiosity.

—John Van DeWalle, Karen Karp, and Jennifer Bay-Williams

Mr. Davis put a lot of thought into his weekly teaching routines and used data from the team's common assessments to determine what students still needed to learn (see story on page 82). Dr. Buck noticed similarities in the strategies Mr. Davis was using with other teachers on the team, as well as teachers he worked with in different schools. Mr. Davis needed to focus on teaching students to make connections between concepts through conceptual understanding and application and spend less time focusing on teaching procedural fluency.

Some students might be able to mimic Mr. Davis's work in class, but many were not making meaning of the mathematics, so they could not apply their learning in later lessons. Additionally, Mr. Davis put an algorithm to real-world problem solving instead of having students read the question and try different solution strategies. The key words *more* and *altogether*, for example, could mean *add*, but might also mean *multiply* or even *subtract*, as in the following examples.

- **Add:** Jarvis has 3 apples. He buys 4 more apples. How many apples does Jarvis have altogether? [3 + 4 = ?]
- **Add or subtract:** Mandy has 3 apples. She buys some more apples. Altogether she has 7 apples. How many apples did Mandy buy? [3 + ? = 7 or 7 − 3 = ?]
- **Add and multiply:** Lola has 3 apples. She buys 5 times more apples at the store. How many apples does Lola have altogether? [3 + (3 × 5) = ?]

Alan H. Schoenfeld (1995), mathematics education researcher, defines *problem solving* as confronting a situation that does not have a ready answer—not merely completing exercises using known procedures. Problem solving offers a question that cannot be immediately answered and focuses more on the process than the solution (Kobett & Karp, 2020).

In defining problem solving, it may also be helpful to define what problem solving is *not*. Problem solving is not demonstrating a series of steps for students and having them practice the procedures repeatedly. When focusing on steps, students may view their current learning as isolated from prior learning and have difficulty applying it to real-life contexts. This approach, focusing on steps and procedures, also positions students as passive learners, assumes they have the necessary background knowledge and skills, and communicates there is only one method to solving a certain type of problem (Van de Walle et al., 2019).

## The *Why* and the *What* of Focusing on Problem Solving

Problem-solving skills are necessary to thrive in the 21st century workplace. Authors John A. Van de Walle, Karen S. Karp, and Jennifer M. Bay-Williams (2019) write, "Skills needed in the 21st century workplace are less about being able to compute and more about being able to design solution strategies" (p. 31). The *Framework for 21st Century Learning*, created by Battelle for Kids (2019), describes the blend between academic skills, social expertise, and the ability to solve problems that arise as necessary to be successful in today's world. Within academic learning, students learn vital academic and life skills, such as critical thinking, problem solving, communication, and collaboration

(Battelle for Kids, 2019). Students need opportunities to question and explore, evaluate and reflect, and apply and extend the mathematics content they are learning to real-world scenarios to navigate their journey through future mathematics courses, as well as their life in an increasingly complicated world.

Problem solving is one of the first process standards described back in 2000 by NCTM in *Principles and Standards for School Mathematics*. The problem-solving process standard highlights the importance of engaging students in meaningful mathematics that lends itself to a variety of entry points and solution strategies. Toncheff and colleagues (2024) define meaningful mathematics as containing "elements that create student agency in learning through reasoning and sense making, while also connecting to students' prior knowledge" (p. 46). Through problem solving, students make sense of the mathematics they are learning and explore applications through reasoning and use of multiple strategies (NCTM, 2000).

Problem solving is also one of the habits learners develop on their journey toward proficiency as described in the standards for mathematical practice (NGA & CCSSO, 2010). The first standard for mathematical practice, "Make sense of problems and persevere in solving them," is a habit of mind needed for students to problem solve (NGA & CCSSO, 2010). This first standard emphasizes the need for students to understand and explain to themselves and others what a problem is asking, consider various approaches to the problem, and monitor their thinking before, during, and after problem solving (NGA & CCSSO, 2010).

In his research, Hattie finds that mathematics problem solving has an effect size of 0.98 (Visible Learning Meta$^X$, 2023). He defines *mathematics problem solving* as teaching mathematical reasoning required to solve practical problems and shares that it has the potential to considerably accelerate student learning (Visible Learning Meta$^X$, 2023). Developing the thinking required for mathematical problem solving transfers to other subject areas and life skills to empower students as learners.

NCTM's (2014) book, *Principles to Actions: Ensuring Mathematical Success for All*, shares eight research-based teaching practices as ways to improve teaching and learning mathematics. The second mathematics teaching practice, "Implement tasks that promote reasoning and problem solving," advocates that teachers and teams use tasks with multiple entry points and invite students to use different strategies, representations, and tools (NCTM, 2014). The strategies students use when solving tasks may be familiar to them from previous experiences, such as creating a picture or visual, using a table or number line, creating and graphing equations, identifying patterns, or constructing similar but simpler problems.

These principles, standards, and practices help ensure all students are engaged in meaningful opportunities to learn mathematics, specifically through problem solving. Van de Walle and colleagues (2019) highlight the following teaching benefits of problem solving:

- Focuses students' attention on ideas and sense making
- Develops mathematical practices and processes
- Develops student confidence and identity
- Builds student strengths
- Allows for extensions and elaborations
- Engages students so there are fewer discipline problems
- Provides formative assessment data
- Invites creativity (pp. 35–37)

Teaching through problem solving engages students in learning mathematics in a relatable context. The traditional approach to teaching mathematics generally involves explaining a concept or skill to students and then having them practice it through abstract problems using only numbers and symbols before solving word problems. Teaching through problem solving is a reverse of the traditional model: Learning begins with a problem or task, and the concepts or skills emerge from it (Van de Walle et al., 2019). Students can tackle the task, even if they do not know how to, with procedural fluency and, in doing so, make sense of the mathematics and connect it to prior learning. Consider the example tasks in figure 6.1 and how each one engages students in solving problems.

The tasks in figure 6.1 allow students to engage in various solution strategies to answer the question. In the first elementary example, students look for patterns when creating a table of solutions, such as one addend increases by one, and the other addend decreases by one.

**Directions:** Identify the set of problems closest to the grade or course you teach. Describe the strategies a student might use to solve the task. Determine which concepts or skills will emerge from the task.

|  | **Example Task** | **What strategies might students use to solve the task?** | **Which concepts or skills emerge from the task?** |
|---|---|---|---|
| **Elementary Examples** | A carton of eggs holds 12 eggs. Some of the eggs are white, and some are brown. How many eggs are white, and how many are brown? | | |
| | Find at least two values for the question mark in the fraction. $$\frac{1}{2} < \frac{2}{?}$$ | | |
| **Secondary Examples** | At the baseball game, Henry bought 2 hotdogs and 6 packs of candy. Hisam bought 1 hotdog and 3 packs of candy. Henry and Hisam spent less than $20 altogether on their snacks. How much is one hotdog? How much is one pack of candy? | | |
| | The perimeter of a scalene triangle is 110 units. If the third side of the triangle is longer than the other two sides, what is the length of the third side? | | |

**Figure 6.1: Examples of problem-solving tasks.**

*Visit **go.SolutionTree.com/MathematicsatWork** for a free blank reproducible version of this figure.*

In the first secondary example, students can use various representations and tools as solution strategies, such as graphing the equations and looking for the point of intersection, creating a pictorial representation of the purchases and looking for common sets of items in each, or using an algebraic approach like substitution.

Implementing tasks as a collaborative team to develop problem-solving skills occurs throughout both Tier 1 and Tier 2 instruction. However, the purpose of the problem-solving task may vary slightly between the two tiers, as shown in table 6.1 (page 86).

Problem solving develops reasoning and sensemaking and provides an opportunity for students to make mathematical connections in their learning. Consider, as a team, how to grow student problem solving most effectively through your instructional practices in Tier 1 and Tier 2.

## The *How* of Focusing on Problem Solving

As shared by educators and authors Kit Norris and Sarah Schuhl (2016), "Problem solving is central to a mathematics classroom" (p. 15). Engage students in learning experiences as often as possible that allow them to apply their prior knowledge while reasoning through new problem situations using contexts that are meaningful and relevant to their lives. Students deserve more than just procedural practice with memorized steps to find the correct answer; they need unique experiences that allow them to both ask and answer *why*.

Table 6.1: Focus on Problem Solving in Tier 1 and Tier 2

| Focus on Problem Solving in Tier 1 and Tier 2 |
| --- |
| • Select problems that allow students to answer, "Why are we learning this?"<br>• Identify structures and routines that support students' perseverance.<br>• Identify success criteria for problem solving and share with students (see chapter 11, page 153, for more information about success criteria).<br>• Engage students in situations that allow them to connect prior knowledge and representations to a new application.<br>• Allow students to share their interpretation of the problem situation.<br>• Allow access to manipulatives, concrete models, and other tools to make sense of the problem.<br>• Engage students in reflecting on their solution strategies before, during, and after problem solving.<br>• Provide opportunities for students to work collaboratively. |

| Focus on Problem Solving in Tier 1 | Focus on Problem Solving in Tier 2 |
| --- | --- |
| • Give students autonomy to choose a strategy of their preference for solving problems.<br>• Have students explain their strategy choice and rationale for using it.<br>• Use problems that focus on development of grade- or course-level skills.<br>• Use problems that may require connecting several pieces of prior knowledge for solving.<br>• Incorporate necessary scaffolds and sequencing in supporting all students' access to problems.<br>• Encourage students to initially generate feedback from their peers during problem solving. | • Encourage students to use a specific strategy or tool (such as manipulatives and technology or digital resources) based on their strengths and background knowledge.<br>• Model the use of specific strategies or tools using a think-aloud before giving students the task as practice.<br>• Give students an initial task containing content below grade level, then work up to grade- or course-level tasks.<br>• Provide additional time as demonstrated by student need.<br>• Provide specific feedback to clarify any student misconceptions or errors. |

The types of tasks used in your lessons determine the level of problem-solving reasoning students are required to do. Mathematics educators Margaret S. Smith and Mary K. Stein (2011, 2018) characterize mathematics tasks as low- or high-level-cognitive-demand tasks (see appendix B, page 179). Having a balance of low- and high-level-cognitive-demand tasks is critical in lesson design and discussed in detail in *Mathematics Instruction and Tasks in a PLC at Work, Second Edition* (Toncheff et al., 2024).

As shared in chapter 3 (page 43), low-level-cognitive-demand tasks are more algorithmic and procedural in nature (see chapter 3 for more information on procedural fluency). High-level-cognitive-demand tasks provide opportunities for students to use multiple strategies and models when solving. They require mathematical reasoning beyond an algorithm or procedure, often deepening conceptual understanding and learning connections for students. High-level-cognitive-demand tasks are most effectively used during instruction with students in groups so they can learn from one another through discourse (see chapter 9, page 129, for more information on classroom discourse).

When your team determines the high-level-cognitive-demand tasks to use during a lesson to develop student reasoning and problem solving, plan for student engagement during the task by asking the following three questions.

1. What are the different models and strategies students might use to solve this task?
2. Some student groups will get stuck. What questions or tools might you use to get students unstuck without lowering the task's cognitive demand?
3. Some student groups will finish early. How might you extend this task in a meaningful way, if needed? (Toncheff et al., 2024)

Prepare in advance to develop problem-solving skills and mathematical reasoning through your intentional use of tasks in the classroom. Use the following strategies in Tier 1 and Tier 2 learning experiences to ensure that all students have access to this level of learning through problem solving.

## Visuals and Tools

Visual representations, diagrams or pictures, and concrete manipulatives and tools provide ways for students to represent mathematical reasoning when solving problems. Graphic organizers, like those used in literacy instruction (for example, KWL chart and What Do You Notice? What Do You Wonder? chart), help students see and make sense of problems (Norris & Schuhl, 2016).

Graphic organizers that emphasize the connection between manipulatives—the part-part-whole relationships in the form of number bonds and tape diagrams—and the number sentence help support elementary students. For middle school and high school students, graphic organizers that emphasize the connection between representations (verbal, graphical, or algebraic) help them understand how situations and relationships can be described using variables.

Students sometimes struggle to understand the relationship between quantities and find it challenging to select appropriate operations that align with the reasoning required to develop problem-solving skills. By using a graphic organizer to understand how the values given in a problem relate to one another, students make better sense of the problem and develop an approach to solving it. Figures 6.2–6.4 (pages 87–88) share graphic organizers meant to help students understand connections between mathematics strategies.

You and your team choose a task and agree on the strategies, objects, or models for students to use when learning through problem-solving graphic organizers. Students rotate to different stations in the room; at each station, they solve using a specific strategy. Students could also stay in their seated groups and complete the graphic organizer. The problem-solving graphic organizer shows students there is more than one way to solve a problem and provides an opportunity for them to choose their favorite strategy for future use.

Figure 6.5 (page 89) shares example tasks using visuals to make sense of a problem. A visual, like a tape diagram, is sometimes considered a graphic organizer because it organizes how a student thinks about numbers and tasks.

Graphic organizers, such as tape diagrams and number bonds, and visuals using pictures or graphs, support students in quantifying information and mapping out the relationships among them. Images like the tape diagram support students in transferring their understanding from a concrete level to an abstract level (She & Harrington, 2022).

In addition to visuals, students can strengthen their understanding of problem solving using concrete models. The graphic organizers in figures 6.2 and 6.3 (page 88) include objects as a strategy to help students solve the task using linking cubes or fraction tiles, respectively.

| Words | Objects |
|---|---|
| James has 7 blocks. He finds 4 more. James needs 20 blocks to build his tower. How many more blocks does James need? | Student shows how to solve the task using linking cubes or blocks while the teacher observes. |
| **Picture** | **Equation** |
| (illustration of blocks) | $7 + 4 + ? = 20$<br>OR<br>$7 + 4 = 11$<br>$20 - 11 = 9$ |
| **Answer:** 9 blocks | |

*Source: Adapted from Norris & Schuhl, 2016.*

**Figure 6.2: Grades K–2 problem-solving graphic organizer.**

*Visit **go.SolutionTree.com/MathematicsatWork** for a free blank reproducible version of this figure.*

**Words**

Emma is making cookies. The recipe calls for $\frac{3}{8}$ cup of brown sugar for 6 giant cookies. Emma wants to make 18 giant cookies. How much brown sugar does Emma need?

**Objects**

Student shows how to solve the task using fraction tiles or a digital number line while the teacher observes.

**Picture**

| $\frac{3}{8}$ cup brown sugar | $\frac{3}{8}$ cup brown sugar | $\frac{3}{8}$ cup brown sugar |
|---|---|---|
| 6 cookies | 6 cookies | 6 cookies |

**Equation**

$18 \div 6 = 3$

$3 \times \frac{3}{8} = \frac{9}{8}$ or $1\frac{1}{8}$

**Answer:** $\frac{9}{8}$ or $1\frac{1}{8}$ cups of brown sugar

**Figure 6.3: Grades 3–5 problem-solving graphic organizer.**

**Words**

Kaelin earned $124 working 8 hours at her job as a hostess. At her second job at the movie theater, she earned $45 working 180 minutes. At which job did Kaelin earn the most per hour?

**Ratio Table**

Hostess

| Money Earned ($) | 124 | 62 | 31 | 15.50 |
|---|---|---|---|---|
| Hours | 8 | 4 | 2 | 1 |

Movie Theater

| Money Earned ($) | 45 | 15 |
|---|---|---|
| Minutes | 180 | 60 |

**Algebraic Representation**

Hostess

$$\frac{\$124}{8 \text{ hours}} = \frac{\$15.50}{1 \text{ hour}}$$

Movie Theater

$$\frac{\$45}{180 \text{ minutes}} = \frac{\$45}{3 \text{ hours}} = \frac{\$15}{1 \text{ hour}}$$

**Tape Diagram**

Hostess

$124

| $15.50 | $15.50 | $15.50 | $15.50 | $15.50 | $15.50 | $15.50 | $15.50 |
|---|---|---|---|---|---|---|---|

Movie Theater

$45

| $15 | $15 | $15 |
|---|---|---|

*Each box shows 1 hour.

**Answer:** Kaelin earns more per hour at the hostess job.

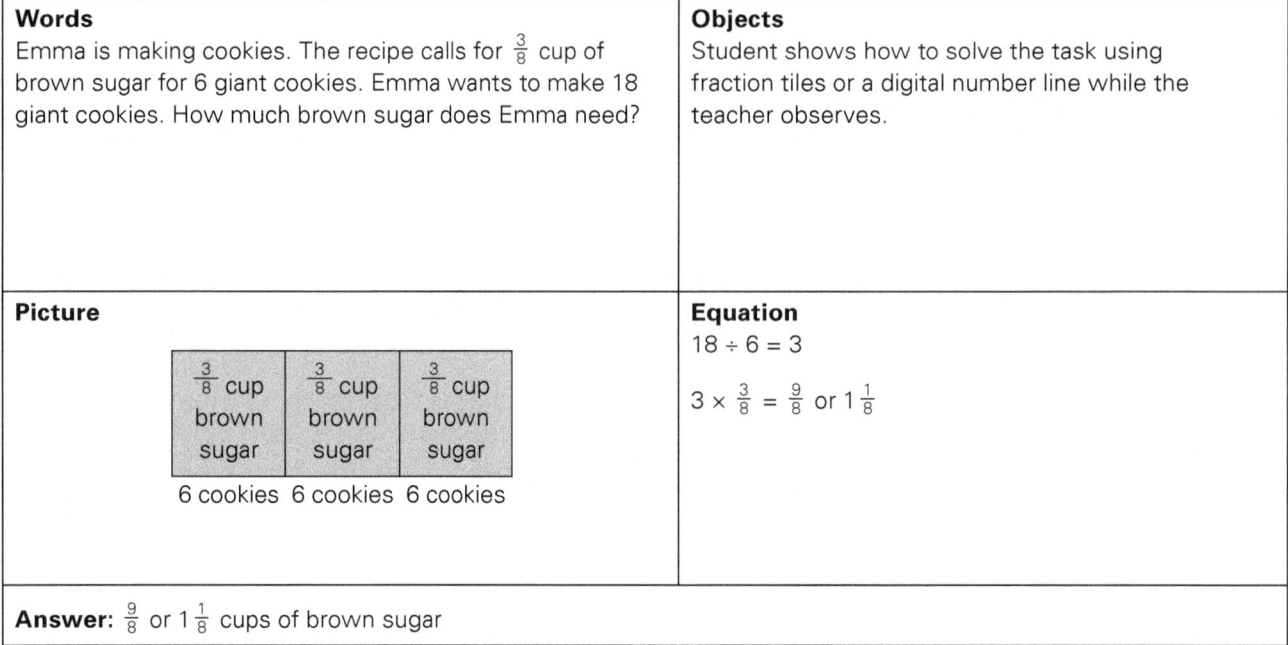

**Figure 6.4: Grade 6 problem-solving graphic organizer.**

*Visit* ***go.SolutionTree.com/MathematicsatWork*** *for a free blank reproducible version of this figure.*

**Directions:** Consider the examples of graphic organizers and visuals used to make meaning of each task. Discuss what other visuals or tools you might use to help students see the mathematics in the task as they grow their problem-solving abilities.

**Grades PreK–K Examples**

Five, Ten, and Twenty Frames

Story Mats

**Grades 1–2 Examples**

Jamie did 13 pushups this morning and did some more pushups in the evening. By the end of the day, Jamie had done 25 pushups. How many pushups did Jamie do in the evening?

Tape Diagram

| 25 pushups | |
|---|---|
| 13 pushups | ? evening pushups |

Number Bond

**Grades 3–5 Example**

Gina studied five days this week, and she studied for the same amount of time each day. Gina studied a total of 275 minutes this week. How many minutes did Gina study each day?

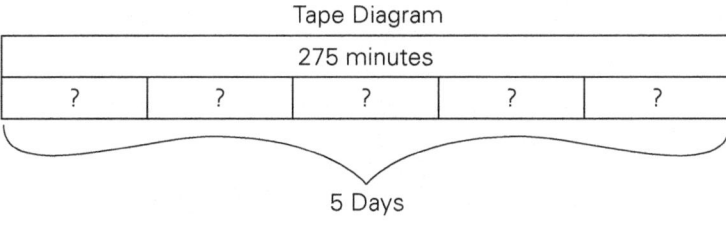 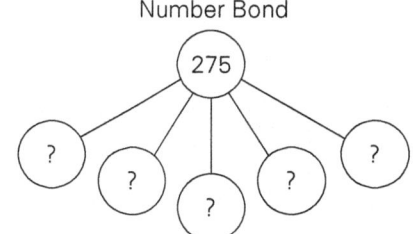

**Grades 6–8 Example**

The ratio of red candies to blue candies in a jar is 4:5. There are 80 red candies in the jar. How many red and blue candies are in the jar altogether?

Ratio of Red to Blue Candies

| R | R | R | R | |
|---|---|---|---|---|
| B | B | B | B | B |

Ratio of Red Candies to Total Candies

| R | R | R | R | | | | | |
|---|---|---|---|---|---|---|---|---|
| T | T | T | T | T | T | T | T | T |

**Figure 6.5:** Examples of graphic organizers and visuals.

continued ▶

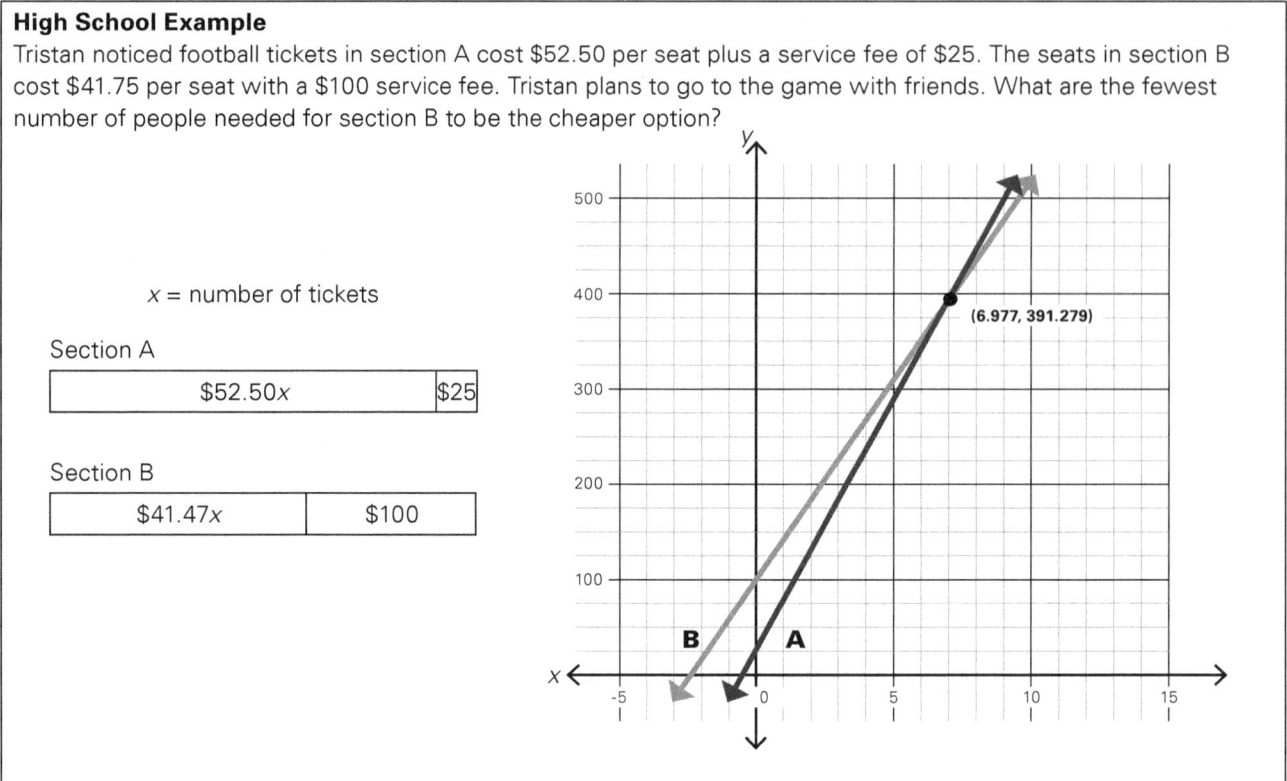

In middle school, students needing support in understanding addition with integers might make sense of the concept using two-color counters as tools. Figure 6.6 shows an example of a problem-solving task requiring the use of two-color counters.

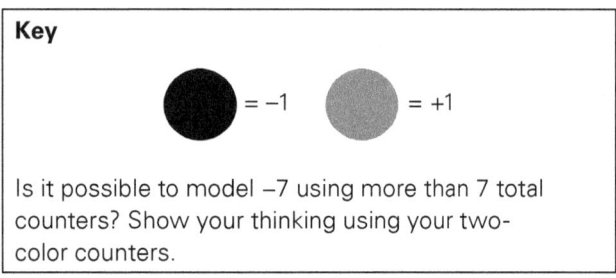

**Figure 6.6: Integer addition task using concrete models for secondary students.**

To answer the question in figure 6.6 and use more than seven counters to model –7 means students will have to explore different combinations of numbers that sum to –7 and understand how zero pairs (one positive counter and one negative counter) can be used to answer the question. They will problem solve ways to create possible solutions. This concrete exploration of two-color counters to create zero pairs helps students develop an understanding of the addition and subtraction properties of equality.

At the high school level, students might solve problems using algebra tiles as tools. Figure 6.7 shows an example of solving an equation using algebra tiles.

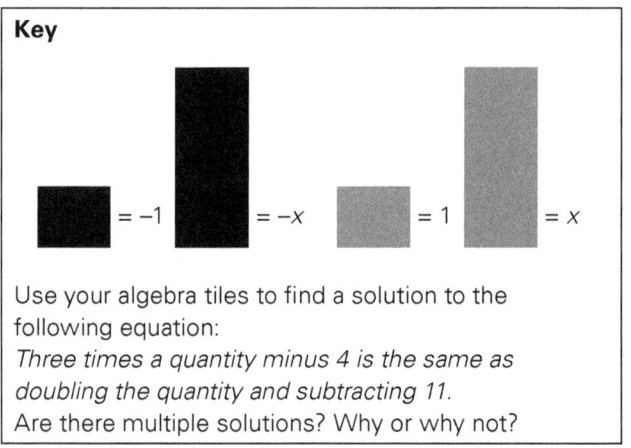

**Figure 6.7: Example of using algebra tiles to solve an equation.**

Using visuals (such as graphic organizers or diagrams) or tools (such as objects and manipulatives) allows students to see connections between what is given in a problem and what they need to find. They visualize their mathematical reasoning and then write their thinking algebraically (using numbers and symbols). See chapter 7 (page 95) for more guidance on using visuals.

## Problems Without Numbers

Word problems are sometimes difficult for students, as the language can become cumbersome and confusing (Gallagher, Ellis, & Weiland, 2021; Kobett & Karp, 2020). One common instructional approach to problem solving is to have students identify key words. However, as shared earlier in this chapter, identfying key words is problematic, as it reduces the depth of mathematics to scanning a problem for a single word and applying an operation associated with that word. This encourages students to do the mathematics removed from its context and fails to consider other meanings a word may have in other content areas or real life (Karp, Bush, & Dougherty, 2014).

Students need to read and make sense of the full task and question. To complicate matters, students also often use the first two numbers they see in a problem and perform an operation, without determining whether they have the right numbers, or operation, to answer the question. Removing the numbers from word problems allows students to focus on understanding the mathematics required to solve the problem rather than collecting any numbers in the problem and applying a random operation.

Consider the problems without numbers examples in figure 6.8. These example tasks have a vague word for the quantity underlined. The tasks might also have only a line where a number should be. Problems without numbers can be used to develop mathematical reasoning abilities needed to effectively solve problems.

Focusing on the operations and relationships between values in a task helps students more effectively read and make sense of word problems and better problem solve with strong mathematical reasoning.

## Three Reads Protocol

Too often, students quickly skim word problems and don't actually read them, or they only read the word problem once. They write down the numbers they see in the problem and apply an operation recently used in class to finish it. Sometimes, the operation does not even match the scenario. When students do this, they are not making sense of the task before planning how to solve it. Students struggling to understand the task or formulate a problem-solving strategy benefit from reading the task closely with different purposes and engaging with the task after each reading. The three reads protocol asks students to make sense of a

**Directions:** Read each task. The underlined word indicates where numbers have been removed from the task. Discuss how to solve each task. What operation or operations are needed?

| Problems Without Numbers Task | Explain How to Solve Each Task |
|---|---|
| **PreK Example**<br>Edward has some grapes. He ate one grape. How many grapes does Edward still have? | Subtract one grape from Edward's total number of grapes. |
| **Grade 1 Example**<br>Thomas filled some buckets with water. Isiah filled more buckets than Thomas. How many more buckets did Isiah fill? | Subtract the number of buckets Thomas filled from the number of buckets Isiah filled.<br>OR<br>Count on from the number of buckets Thomas filled until reaching the number Isiah filled. Determine how many numbers were used when counting on. |
| **Grade 5 Example**<br>Sarai's soccer team played some games this season, and Sarai started in some (fraction) of the games. How many games did Sarai start in? | Multiply the total number of games by the fraction of the games Sarai played. |
| **Algebra or Integrated Mathematics 1 Example**<br>Anna's plant started at a height. It grew some inches over a period of time at a constant rate. How long will it take for her plant to grow to a specified height? | Create a linear equation with the y-intercept as the original height of the plant. The slope is the rate showing the number of inches the plant grew in a period of time.<br>Set the linear equation equal to the specified height and solve algebraically or with a graph. |

**Figure 6.8: Sample problems without numbers.**

*Visit* **go.SolutionTree.com/MathematicsatWork** *for a free blank reproducible version of this figure.*

mathematics word problem by reading the problem three times with a different focus for each read, a variation of close reading in literacy.

There are several variations of the three reads protocol. Sarah Schuhl (2023) gives the following directions for each read through, with students working in pairs.

1. Students read a task independently the first time, aiming to understand the context of the task. They answer the question, "What is the story?" For example, is the problem about skiing, going to the grocery store, or driving a car? Ask partner A to tell partner B what the task is about.

2. Students read the task a second time to understand what the task is asking. They answer the question, "What is the question we are trying to answer?" Ask partner B to tell partner A the question they are answering.

3. Students read the task a third time to answer the question, "What do I know about the problem, and what do I need to know?" or "What can we predict is the answer?" At this point, partner A and partner B work together to answer the question posed.

After reading the task three times and understanding it, students determine a plan to use the information given in the problem and reason toward a solution. The reasoning used throughout the three reads protocol and the work leading to a solution grows students' problem-solving abilities.

Another version of the three reads protocol connects to the problems without numbers strategy, as shown next.

1. **First read:** Read to understand the story. What is the problem about? If possible, remove the numbers to reinforce sensemaking.

2. **Second read:** Read to understand the mathematics. Add the numbers back into the problem and analyze the language used. What information is given? What don't you understand? What model can you create to represent the story?

3. **Third read:** Plan how to solve. How can you solve the problem?

Figure 6.9 includes elementary and secondary examples of anchor charts used to teach and reinforce the use of mathematical language using the three reads protocol.

As a team, determine how to use the three reads protocol, understanding that the first read is for context, whether or not the task includes numbers. The remaining two reads and work help students understand the question, identify the information given and needed, and work to find a solution. You may also ask students to predict an answer to develop their number sense and ability to assess an answer's reasonableness. The main purpose of using three reads is for students to internalize and apply this routine on their own whenever they encounter word problems.

## Conclusion

Engaging students in problem solving allows them to make meaning of the mathematics they are learning in the classroom. When students are solving problems, they know they can use previously learned strategies and models to tackle a task.

Encourage students to approach problems in several ways, and be cautious about taking over student thinking. To assist them in exploring and communicating their strategies, students use visuals and tools to make their thinking visible and to better understand the task. Students learn through deep reasoning and connections inherent to solving problems. The tasks your team chooses provide opportunities for meaningful problem solving to occur in the classroom.

## Questions to Consider for Next Steps

As a collaborative team, use the following questions to reflect on your current practices and determine any next steps for building number sense.

1. How does your team choose which mathematical tasks to use to develop students' problem-solving skills?

2. How does your team identify and use high-level-cognitive-demand tasks in the classroom to solve problems?

3. What pictures, tools, strategies, representations, or models do your students use when problem solving?

4. How might you teach students to pause and make sense of a task before beginning to solve it?

5. What is a strategy your team wants to try related to deepening students' problem-solving skills in your current or next unit?

**Elementary Example**

## Three Reads Protocol

1st read: Mateo's garden has ___ rows of plants with ___ plants in each row.

2nd read: Mateo's garden has _2_ rows of plants with _7_ plants in each row.

Row 1 → 🌱 🌱 🌱 🌱 🌱 🌱 🌱
Row 2 → 🌱 🌱 🌱 🌱 🌱 🌱 🌱

3rd read: Mateo's garden has 2 rows of 7 plants in each row. **How many plants are in Mateo's garden?**

$2 \times 7 = 14$
2 rows of 7 is 14
2 equal groups of 7 is 14

Mateo has _14_ plants in his garden.

*Thought bubbles:* Skip the numbers. Skip the question. Make sense of the problem. Include the numbers. Draw a picture. Read the whole problem, including the question. What is the question asking? How do you know?

**Secondary Example**

## Three Reads Protocol

1st read: An Uber ride in Gotham City is _____ for the first _____ mile and an additional mileage charge of _____ for each additional _____ mile.
 *Skip the numbers.   *Skip the question.   *Make sense of the problem.

2nd read: An Uber ride in Gotham City is _$2.40_ for the first _$\frac{1}{2}$_ mile and an additional mileage charge of _$0.20_ for each additional _0.1 mile_.
 *Include the numbers.   *Draw a picture.

$\frac{1}{2}$ mile | 0.1 mile
$2.40 | $0.20

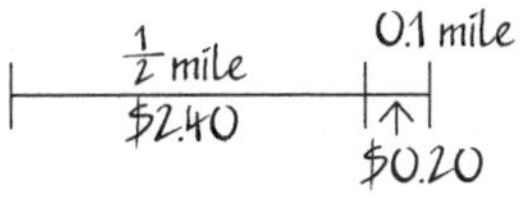

3rd read: An Uber ride in Gotham City is $2.40 for the first $\frac{1}{2}$ mile and an additional mileage charge of $0.20 for each additional 0.1 mile. **How far can you ride for $10?**
 *Read a third time.   *What is the question asking?   *How do you know?

*Source: © 2024 by Georgina Rivera. Used with permission.*

**Figure 6.9: Three reads protocol anchor chart examples.**

> It is not uncommon for mathematics educators to create a goal to ensure all students are procedurally fluent. In her earlier years of coaching elementary mathematics, Jenn remembers supporting teams at her school in Virginia to reach a fluency goal but could not immediately find the path to get there. Many students struggled to make sense of the four operations with whole numbers, fractions, and decimals. So, attempting to erase common misconceptions, Jenn guided teams to help students develop mathematical understanding by using tools to model operations, making connections to place value understandings, and drawing pictorial representations when computing. While the number of students who were misusing algorithms and procedures decreased, the number of students with inefficient strategies increased. This inefficiency led to a major gap in access to grade-level instruction.
>
> Students were computing accurately, and even thinking flexibly, while using strategies that made sense to them. However, many learners stalled in their progress toward fluency and relied too heavily on tools and drawings to successfully compute. Jenn and her team needed to find a bridge between a conceptual understanding of the operations and procedurally fluent strategies that could be applied to grade-level problems. While the foundation for students developing procedural fluency was in place, teachers were on a mission to find the appropriate next steps for students becoming mathematically fluent.

# CHAPTER 7

# Develop Procedural Fluency

> Flexibility with a variety of computational strategies is an important tool for a mathematically literate citizen to be successful in daily life.
>
> —John Van de Walle

Fluency with mathematics facts and procedures has long been a roadblock for students to learn grade- or course-level mathematics. Students sometimes have difficulty developing fluency because they do not yet understand the concept they're learning, as was the case when Jenn began working with the teams in her school (see story on page 94). Other times, students understand the concept, but they are not yet efficiently and accurately working with numbers and procedures. Jenn and her colleagues found it difficult to get some students to develop procedural fluency because they were successful with less efficient strategies.

It can be a challenge to teach higher-order reasoning when students cannot yet demonstrate efficiency and accuracy with foundational facts and procedures. While it may be a challenge, it is not impossible. Students can learn grade- or course-level mathematics while still working to develop mathematical fluency, just as they can learn grade-level reading reasoning while still learning to read independently. Fluency cannot just be expected; it must be taught.

## The *Why* and the *What* of Developing Procedural Fluency

NCTM (2023) defines *procedural fluency* as follows:

> The ability to apply procedures efficiently, flexibly, and accurately; to transfer procedures to different problems and contexts; to build or modify procedures from other procedures; and to recognize when one strategy or procedure is more appropriate to apply than another.

This definition does not mention speed but rather focuses on efficient strategies and accuracy with flexible thinking. Timed tests from kindergarten through third grade are not equivalent to procedural fluency, and procedural fluency is more than quick recall of basic facts with the four operations. For example, in grades 3–5, students develop procedural fluency for fraction and decimal operations; in grades 6–8, students develop procedural fluency to solve proportions, find precents, and graph linear equations; and in high school, students develop procedural fluency to graph functions and transformations. Some key fact fluency and procedural fluency skills are listed in table 7.1 (page 96), along with the grade when fluency is expected.

Students must think flexibly, make decisions about which strategies are most effective, and apply their understanding to a wide range of problem settings to be considered fluent.

Procedural fluency with the standard algorithms for addition and subtraction is expected in fourth grade, multiplication in fifth grade, and division in sixth grade. Students begin working with operations well before the grade when fluency with a standard algorithm is expected. In *Elementary and Middle School Mathematics: Teaching Developmentally* (2019), mathematics educators John A. Van de Walle, Karen S. Karp, and Jennifer M. Bay-Williams write:

> Two things to remember in teaching the standard algorithm for addition:
> 1. Be sure students continue to view it as one possible algorithm that is a good choice in some situations (just as invented strategies are good choices in some situations), and

Table 7.1: Key Fact and Procedural Fluency Expectations by Grade Level

| Grade Level | Fluency Standard |
|---|---|
| PreK | • Count to 10. |
| Kindergarten | • Count to 100 by ones and by tens. (K.CC.A.1)<br>• Fluently add and subtract within 5. (K.OA.A.5) |
| Grade 1 | • Add and subtract within 20 using a variety of strategies. (1.OA.C.6)<br>• Fluently add and subtract within 10. (1.OA.C.6) |
| Grade 2 | • Fluently add and subtract within 20 using mental strategies. By end of grade 2, know from memory all sums of two one-digit numbers. (2.OA.B.2) |
| Grade 3 | • Fluently add and subtract within 1,000 using strategies and algorithms based on place value, properties of operations, and the relationship between addition and subtraction. (3.NBT.A.2)<br>• Fluently multiply and divide within 100, using various strategies. (3.OA.C.7)<br>• Find the area of a rectangle using the formula $A = b \times h$. (3.MD.C.7a) |
| Grade 4 | • Fluently add and subtract multidigit whole numbers using the standard algorithm. (4.NBT.B.4)<br>• Add and subtract mixed numbers with like denominators. (4.NF.B.3c) |
| Grade 5 | • Fluently multiply multidigit whole numbers using the standard algorithm. (5.NBT.B.5)<br>• Add and subtract fractions with unlike denominators. (5.NF.A.1)<br>• Use formulas for the volume of right rectangular prisms ($V = l \times w \times h$ or $V = B \times h$) to solve problems. (5.MD.C.5b) |
| Grade 6 | • Fluently divide multidigit numbers using the standard algorithm. (6.NS.B.2)<br>• Fluently add, subtract, multiply, and divide multidigit decimals using the standard algorithm for each operation. (6.NS.B.3)<br>• Solve equations of the form $x + p = q$ or $px = q$. (6.EE.B.7)<br>• Use the formulas for areas of triangles and special quadrilaterals to solve problems. (6.G.A.1) |
| Grade 7 | • Add, subtract, multiply, and divide rational numbers. (7.NS.A.1 & 7.NS.A.2)<br>• Solve equations of the form $px + q = r$ or $p(x + q) = r$. (7.EE.B.4)<br>• Use proportions to solve problems. (7.RP.A.3)<br>• Use the formulas for area and circumference of a circle to solve problems. (7.G.B.4) |
| Grade 8 | • Perform operations with numbers expressed in scientific notation. (8.EE.A.4)<br>• Solve one-variable linear equations, including rational coefficients and constants. (8.EE.C.7)<br>• Graph linear equations. (8.F.B.4)<br>• Solve problems involving the volume of cones, cylinders, and spheres. (8.G.C.9) |
| High School | • Solve quadratic equations in one variable. (HSA-REI.B.4)<br>• Solve systems of linear equations exactly and approximately. (HSA-REI.C.6)<br>• Solve literal equations. (HSA-CED.A.4)<br>• Graph functions using key features. (HSF-IF.C.7)<br>• Find the zeros of functions. (HSF-IF.C.8)<br>• Graph transformations of a function. (HSF-BF.B.3) |

*Source for standards: NGA & CCSSO, 2010.*

*Visit **go.SolutionTree.com/MathematicsatWork** for a free reproducible version of this table.*

2. As with any procedure (algorithm), it must begin with the concrete, and then explicit connections must be made between the concept (in this case, regrouping) and the procedure. (p. 261)

Beyond the standard algorithms for addition, subtraction, multiplication, and division are other algorithms (sets of rules or steps) for procedural fluency. Students in second grade might add 45 + 31 by adding the tens and then the ones (4 tens + 3 tens = 7 tens and 5 ones + 1 one = 6 ones, so the answer is 7 tens and 6 ones or 76). Adding from larger place value to smaller place value and regrouping as needed is an algorithm—though not the standard algorithm—and one that often helps students with mental computations. In middle school, students learn to apply formulas for areas, surface areas, and volumes of special shapes. In high school, the main algorithm for graphing a linear equation given in standard form ($Ax + By = C$) is to find and plot the $x$-intercept and $y$-intercept and connect the two points to form a line. However, a student could also rewrite the linear equation into slope-intercept form ($y = mx + b$) and then use an algorithm to graph the line using the $y$-intercept and slope.

Consider how to teach algorithms and procedures by building conceptual understanding for the skill with concrete objects and pictures, developing meaning through application, and then helping students make connections between the concept and the algorithm, understanding that the algorithm is one way to demonstrate fluency (see more information about teaching rigorous mathematics in chapter 1, page 9).

Delaying the teaching of procedures gives students time to make sense of concepts while connecting models to invented strategies, and it allows for intentional instruction to happen so students understand why and how algorithms work. In some cases, it helps students compute more efficiently (NCTM, 2014; Smith et al., 2017).

Students' mathematics learning typically starts with concrete objects, transitions to pictorially representing strategies, and ends with students showing their mathematics thinking more abstractly using numbers and mathematical symbols. The acronym *CRA*—concrete, representational (sometimes called semi-concrete), and abstract—stems from the early work of Jerome S. Bruner and Helen J. Kenney (1965), who describe learning through the stages of action (using concrete objects), pictures, and abstract symbols (numbers and mathematical symbols).

### CRA Method in Tier 1

The CRA method is how students initially learn mathematics in Tier 1. These three types of thinking—concrete, representational, and abstract—allow students to have a variety of solution strategies when solving problems and understand any procedural fluency algorithm as one way to solve a problem. For example, consider composing and decomposing numbers in the primary grades. First, students use manipulatives (like linking cubes), then pictures (ten frames or dots), and finally, students write addition and subtraction equations. Students more deeply understand mathematics as they strengthen their concrete thinking and link it to pictures. As they then make connections between the concrete models, pictures, and abstract numbers and symbols, they more fully learn concepts. Kindergartners do not stop using manipulatives when they learn to draw pictures and write numbers. Instead, they merge the three concepts to make sense of mathematics.

As students explore operations with fractions in grades 3–5, they might use fraction bars; then number lines, shapes, and tape diagrams; and then connect the representations to their understanding of operations with fractions and fraction equivalence using numbers and symbols. Students in middle and high school might explore two-dimensional geometry using a geoboard or three-dimensional geometry using objects, drawing pictures, and finally using the pictures to solve mathematical problems using numbers and symbols.

### CRA Method in Tier 2

CRA is also a recommendation for Tier 2 interventions (Fuchs et al., 2021). According to Fuchs and colleagues (2021), the U.S. Department of Education's recommendations for CRA include the following:

> Fade out concrete and semi-concrete representations as students become accurate with doing the work abstractly [and work together to have] consistency in the types of representations shared in core classroom instruction and during intervention sessions, throughout the year and across

grades, [which] is critical. Consistent use is particularly important for students who are struggling to grasp a concept or operation. (pp. 27–28)

Again, though procedural fluency may not include concrete models and pictures, use both routinely with abstract numbers and symbols in Tier 2 to develop meaning for procedural fluency so students learn other efficient ways to solve problems.

For CRA to be effective, give students opportunities to explore concepts using tools and concrete models so learning is interactive and students see how to apply their thinking. By representing problems using drawings, number lines, and other pictorial models, students make sense of the tools they were just using to explore the concept. However, pictures are not enough, so it is important to make connections between the concrete and pictorial to transfer these methods into more abstract representations (mathematics numbers and symbols), leading to procedural fluency.

Moving to procedures too quickly (if place value and numeracy concepts are not embedded in learning), especially in the elementary grades, can lead to students forming misconceptions and retaining less of what they learn, which contributes to mathematics anxiety (NCTM, 2023). Author Robert J. Wright and colleagues (2015) explain, "The formal algorithms are still regarded as important, but they are taught after children have developed facile, informal, mental strategies" (p. 31). If students only know how to solve problems using algorithms, consider having them also draw pictures and use models so they develop a deeper understanding of mathematics.

In *Adding It Up: Helping Children Learn Mathematics* (Kilpatrick et al., 2001), the National Research Council determines mathematical proficiency relies on five strands, one of which is procedural fluency. The authors state, "Without sufficient procedural fluency, students have trouble deepening their understanding of mathematical ideas or solving mathematics problems. The attention they devote to working out results they should recall or compute easily prevents them from seeing important relationships" (p. 122). Being procedurally fluent means students make stronger connections between types of numbers or operations and mathematical concepts. It means students have accurate and efficient methods for solving proportions and equations and graphing functions.

Procedural fluency is an aspect of mathematics taught in Tier 1 core instruction. However, teachers address procedural fluency in Tier 2 interventions when students are not procedurally fluent and should be. Table 7.2 outlines some keys to building procedural fluency in Tier 1 and Tier 2 instruction.

Working toward procedural fluency is a critical part of mathematics in Tier 1 instruction, but it is not the entire goal. Remember, rigor in mathematics is defined as a balance of conceptual understanding, application, and procedural fluency (see more about teaching rigorous mathematics in chapter 1, page 9; NGA & CCSSO, 2010). Once students develop procedural fluency in the expected grade or course, they can apply it to future content and use it to make sense of the mathematics they are learning.

Procedural fluency becomes a focus for Tier 2 instruction as an intervention when students are still working to develop efficient, accurate, and flexible thinking and procedures past the expected mastery level. One specific type of procedural fluency is fact fluency, meaning students add, subtract, multiply, and divide fluently within 20 and 100, respectively. As with number sense (see chapter 5, page 69, for more information about number sense), students still working to learn fact fluency in middle and high school need Tier 3 support. In fact, secondary students who have not yet learned their facts are most likely simultaneously struggling with procedural fluency and number sense.

To address fact fluency as a team, consider using number talks (see chapter 5, page 69) for five to ten minutes per day before teaching the grade- or course-level lesson during the rest of the instructional block. Use small groups during part of instruction or play fact fluency games, such as Math-O (Bingo board with answers, and you provide the fact fluency prompts). Students needing Tier 3 intensive reinforcements also need Tier 1 grade-level instruction and Tier 2 intervention on grade-level essential standards. The next section shares strategies for how to support students developing procedural fluency, with a focus on Tier 1 and Tier 2 instruction.

Table 7.2: Develop Procedural Fluency in Tier 1 and Tier 2

| Develop Procedural Fluency in Tier 1 and Tier 2 ||
|---|---|
| <ul><li>Teach conceptual understanding and use it to develop procedural fluency.</li><li>Teach using the CRA method with models and tools to support student thinking when learning how to compute and when making sense of strategies.</li><li>Determine consistent manipulatives to use in each unit across your collaborative team.</li><li>Use word problems to provide context for a concept (application), and have students use manipulatives, draw a picture, and write their thinking abstractly using numbers and symbols.</li><li>Have students share different ways to solve tasks and analyze the strategies used to determine which are most efficient and effective.</li><li>Intentionally connect the work of operating with whole numbers to operating with fractions, decimals, and integers.</li><li>Read equations (numbers and symbols) to understand each one's meaning.</li><li>Routinely engage in subitizing to see values and equal groups for early fluency with addition and subtraction.</li></ul> ||
| **Develop Procedural Fluency in Tier 1** | **Develop Procedural Fluency in Tier 2** |
| <ul><li>Determine consistent strategies to model and highlight in each unit across your collaborative team.</li><li>Allow the use of multiple strategies, sometimes student invented, to build flexible thinkers rather than mimic rote procedures.</li><li>Routinely practice procedural fluency for short times during instruction.</li></ul> | <ul><li>Ensure that working with concrete models and drawings during re-engagement coincides with teaching abstract representations of operating with numbers.</li><li>Revisit and emphasize the CRA method to gain an understanding of students' early use of algorithms and procedures.</li><li>Focus on the conceptual understanding part of the skill being learned.</li></ul> |

## The *How* of Developing Procedural Fluency

There are many ways to develop procedural fluency through Tier 1 and Tier 2 instruction. The strategies shared in this chapter support the CRA framework, as well as help students move toward efficient and flexible strategies and procedures when learning future mathematics concepts. This section's goal is not to provide an exhaustive list of strategies, but rather show possibilities that connect with your classroom's existing practices for you and your team to discuss and implement.

Students needing Tier 2 support with procedural fluency may become overwhelmed by the sheer number of ways problems can be solved. Explicit guided instruction regarding a limited set of strategies that connect very concretely may be a more reasonable approach for Tier 2 learning experiences. As a team, determine the desired procedural fluency for various skills to identify the most effective models and strategies to use to develop that procedural fluency.

## Flexible Strategies Versus Memorization

Developing fact fluency in preK–3 is often considered essential for students to reason through new concepts without simultaneously trying to make sense of basic mathematics facts using the four operations. To establish lasting success, team members discuss and understand how students gain fluency with basic facts over time and the progression of that learning. In *Catalyzing Change in Early Childhood and Elementary Mathematics*, NCTM (2020) shares five critical ideas of whole-number mastery that students must understand for fact fluency:

1. Dimensions of number sense and operation sense that foster flexibility and confidence in children's numerical reasoning
2. Importance of subitizing across all grade levels, PreK–12 for developing knowledge of quantitative relationships
3. Deeper understanding of place value concepts that support meaningful work with operations

4. Learning basic number combinations through sense making, not memorization
5. Moving from additive to multiplicative thinking as a major milestone of mathematical reasoning (pp. 78–79)

Figure 7.1 shows a learning progression that allows students to work toward immediate recall and have strategies to fall back on when they have not yet memorized facts.

Emphasize noncounting strategies as students become more fluent. The goal is for students to immediately recall basic facts, but if they cannot, they can select a noncounting strategy, use patterns, or pull from their understanding of properties to derive the basic fact.

### Known Ten Strategy

Engaging students in routines that involve visuals and tools, so they learn how to use known facts such as ten combinations, ensures students have a pathway for moving away from counting by ones. Utilizing counters on a double ten frame requires students to engage in composing and grouping the same sum in multiple ways. Not only does a double ten frame serve

| Subitizing Images | Addition and Subtraction Fact Fluency | Multiplication and Division Fact Fluency |
|---|---|---|
| Single Images (Small Numbers) 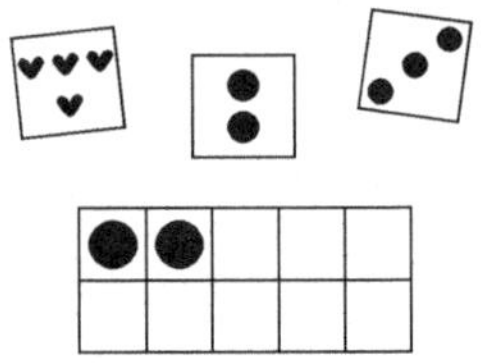 | Counts All 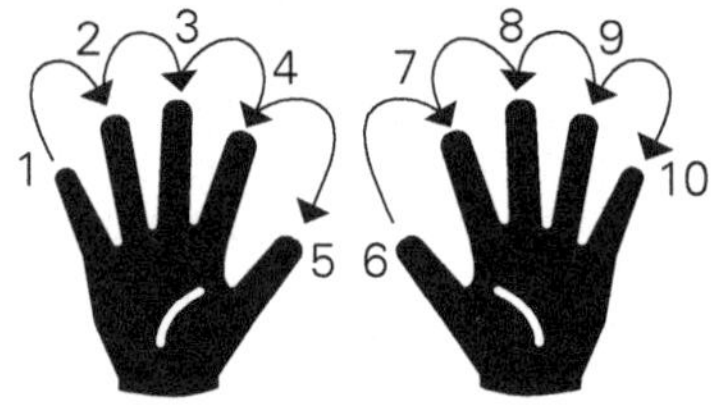 | Counts All 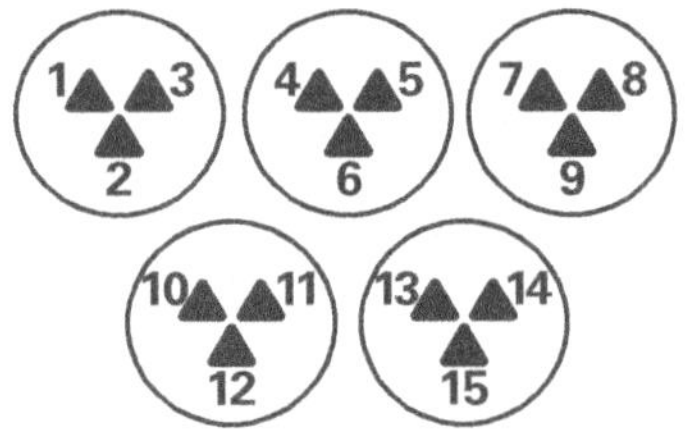 |
| Single Images (Up to Ten) 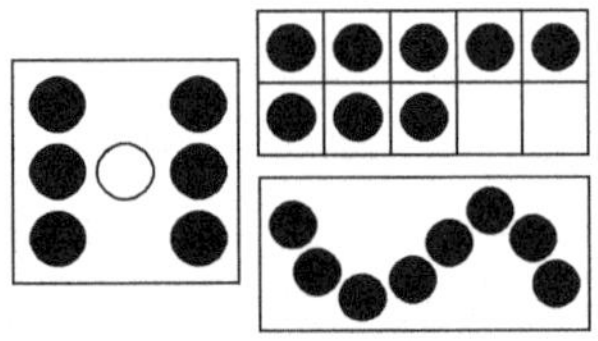 | Counts On or Back 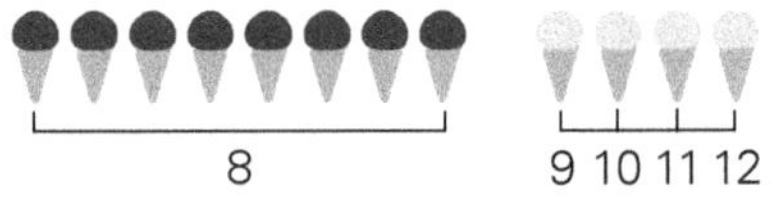<br>8 + 4 = 12 | Skip Counts or Uses Repeated Addition 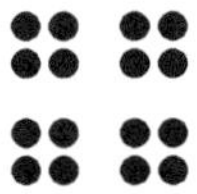<br>4 + 4 + 4 + 4 = 16<br>4, 8, 12, 16 |
| Two Images (Up to Twenty) 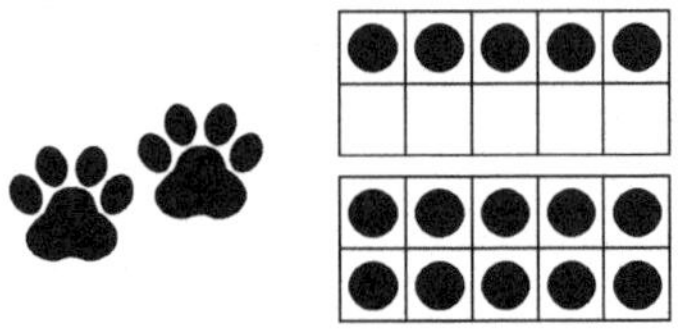 | Uses a Known Fact<br>I know 3 + 4 is 7 because it is like **3 + 3 and 1 more**.<br>**15 − 7 = 8**<br>Use Make a Ten strategy:<br>15 − 5 = 10<br>10 − 2 = 8 | Uses a Known Fact<br>Uses knowledge of foundational 2s, 5s, and 10s to solve other facts. 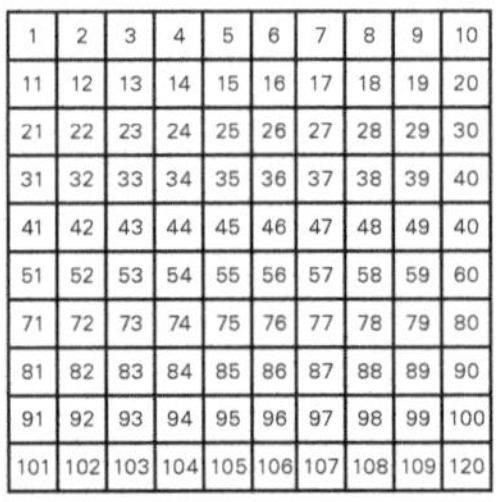 |
| Base Ten and Multiple Groups 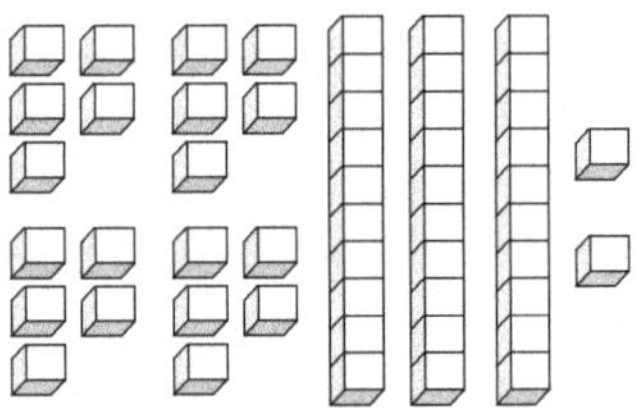 | Immediate Recall <br>I know all the doubles facts!<br>I can say all the numbers that make 10! | Immediate Recall<br>Knows multiplication facts and applies that knowledge to related division facts. |

**Figure 7.1: Fact fluency progression of strategies.**

as a concrete model of the associative property, but it also helps students move away from counting forward or backward and instead creates a known ten to get a total more efficiently.

Figure 7.2 shows an example of how teachers engage students in routines that start with concrete models and objects then transition to the matching abstract equations for making a known ten. This creates flexible strategies to make a known ten without simply memorizing combinations.

You can also bridge to other important concepts, such as equality, when working with students to find a known ten. Combining 9 and 3 is the same as finding the sum of 10 and 2. Assess students' knowledge and readiness for making a known ten and their ability to understand equality while they are solving problems, playing games, and engaging in number talks that encourage sensemaking and sharing how they arrive at an answer.

## Vertical Progression of Strategies

In addition to supporting student thinking, communicating vertical pathways toward fact and procedural fluency through the grades helps align teaching practices for you, your team, and your school. Shared learning in CRA progressions strengthens teacher understanding and develops a common approach to fluency that benefits learners.

Consider how to create visual examples of CRA strategies to develop procedural fluency for specific concepts, as shown in figure 7.3 (page 102). Students interchangeably interact with concrete models, pictures, and abstract equations and symbols as they make connections and deepen their conceptual understanding. Creating visual examples for developing procedural fluency also supports learning at home.

Another example of connecting strategies vertically from one grade to the next happened at Jenn's elementary school, which was mentioned in the scenario at the beginning of this chapter (page 94). Her story shows how learning progresses through the grades, and teams may need clarity on their role in the learning, especially when students have misconceptions.

> When Jenn was coaching teams, she realized students were confusing notations in their drawings when they exchanged a ten for 10 ones and when they subtracted 10. This misconception led to a schoolwide effort to clarify

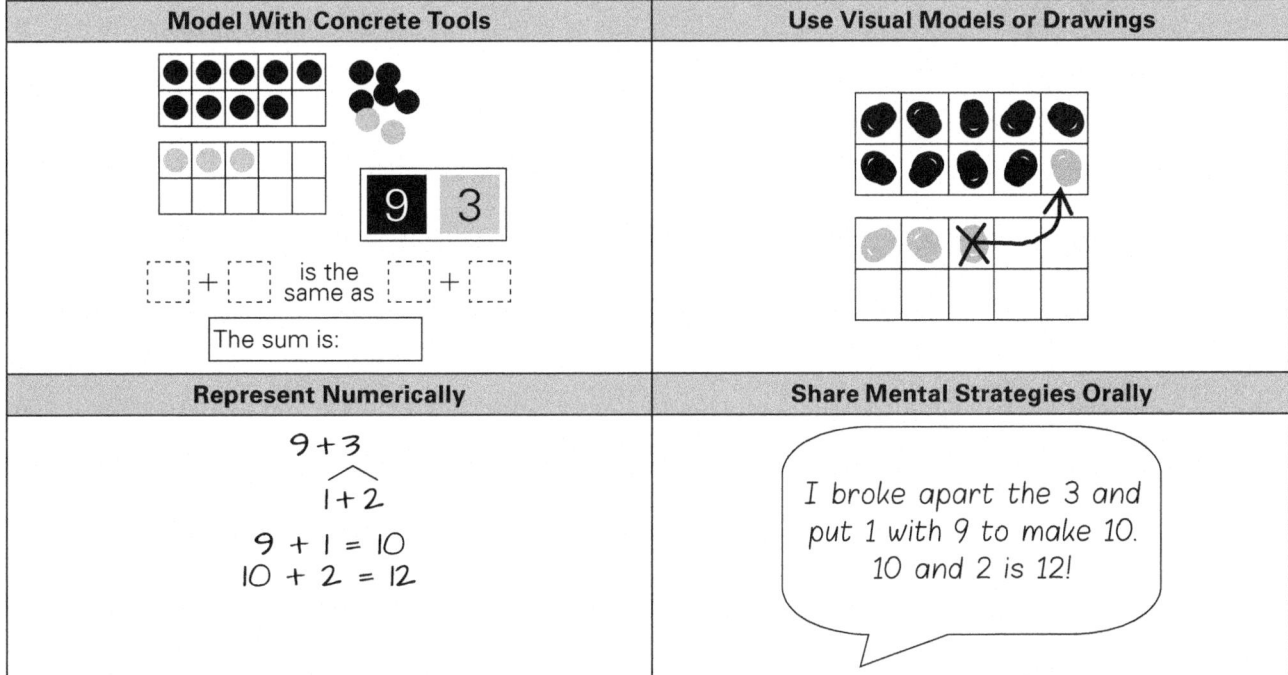

**Figure 7.2: Known Ten strategy from concrete to abstract.**

## Subtraction With Regrouping: From Concrete Tools to Numbers and Algorithms
### 42 − 18

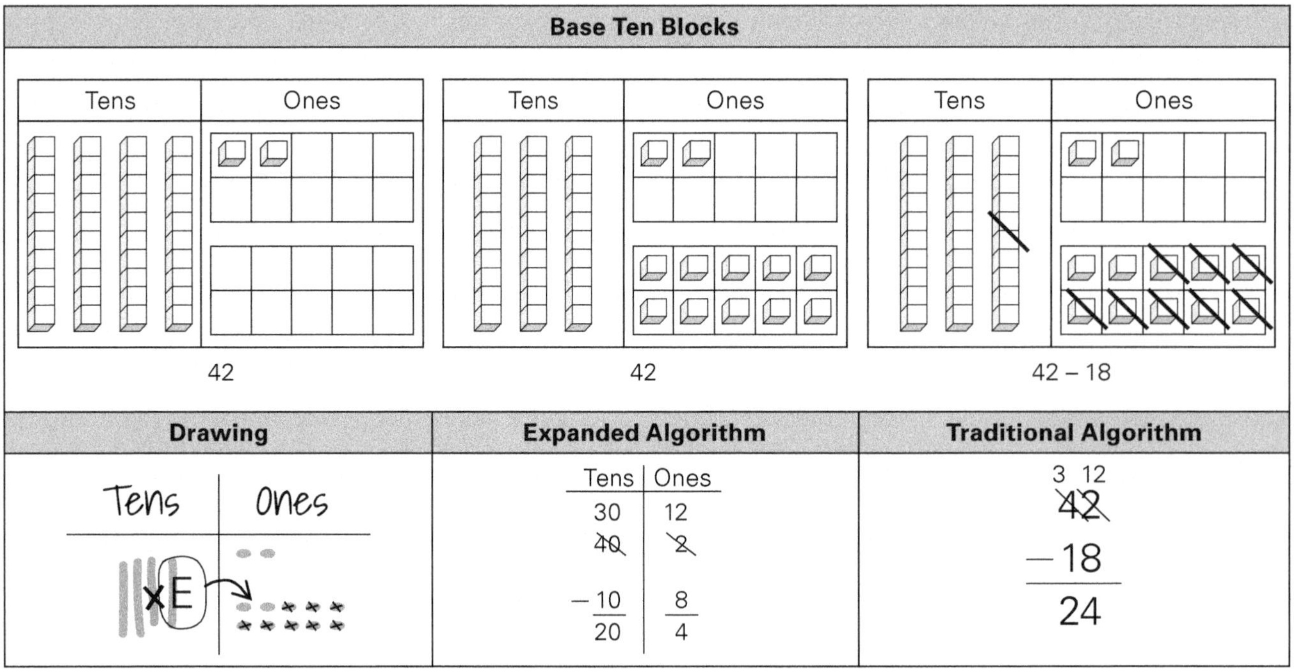

| Base Ten Blocks | | |
|---|---|---|
| Tens \| Ones <br> 42 | Tens \| Ones <br> 42 | Tens \| Ones <br> 42 − 18 |
| **Drawing** | **Expanded Algorithm** | **Traditional Algorithm** |
| Tens \| Ones | Tens \| Ones <br> 30 \| 12 <br> 4̶0̶ \| 2̶ <br> −10 \| 8 <br> 20 \| 4 | 3 12 <br> 4̶2̶ <br> −18 <br> 24 |

Figure 7.3: Multidigit subtraction progression of strategies.

Visit **go.SolutionTree.com/MathematicsatWork** *for a free reproducible version of this figure.*

exchanging a ten in drawings with an E for exchange rather than crossing out the ten. Instead of addressing the misconception with just one grade, Jenn and her team worked with all the grade-level teams (vertical teams). Jenn discussed the need to find a middle ground between drawing pictures and using the standard algorithm with each team.

The expanded algorithm allowed teachers and students to work with and name full values for digits based on their place value rather than single digits (for example, using 30 and 4 for 34 instead of 3 and 4). Using full values connected more directly to the models and drawings students worked with. The vertically shared CRA process allowed students to seamlessly move from one grade level to the next with similar recording experiences. More students schoolwide began using procedurally fluent strategies.

One way students develop procedural fluency vertically at the secondary level is by solving equations. They use various strategies as they solve one-step equations in sixth grade, two-step equations in seventh grade, and multistep equations in eighth grade through high school.

Two examples of concrete models to use when developing procedural fluency with solving equations are algebra tiles or a balance scale with one object representing the variable and a different object representing the number 1. Gather the correct number of variable objects and number objects to match the equation and place them on a balance scale. As students add, subtract, multiply, or divide using properties of equality to solve the equation, they move objects to keep the scale balanced until the equation is solved. After such a concrete representation, students might draw a model and eventually solve the equation algebraically for procedural fluency.

Figure 7.4 shows some examples of how to solve an equation using concrete objects, pictures, and algorithms.

In high school, procedural fluency with graphing functions involves students graphing key features using knowledge about the coefficients for the function's variables. For example, in $y = mx + b$, the $m$ represents

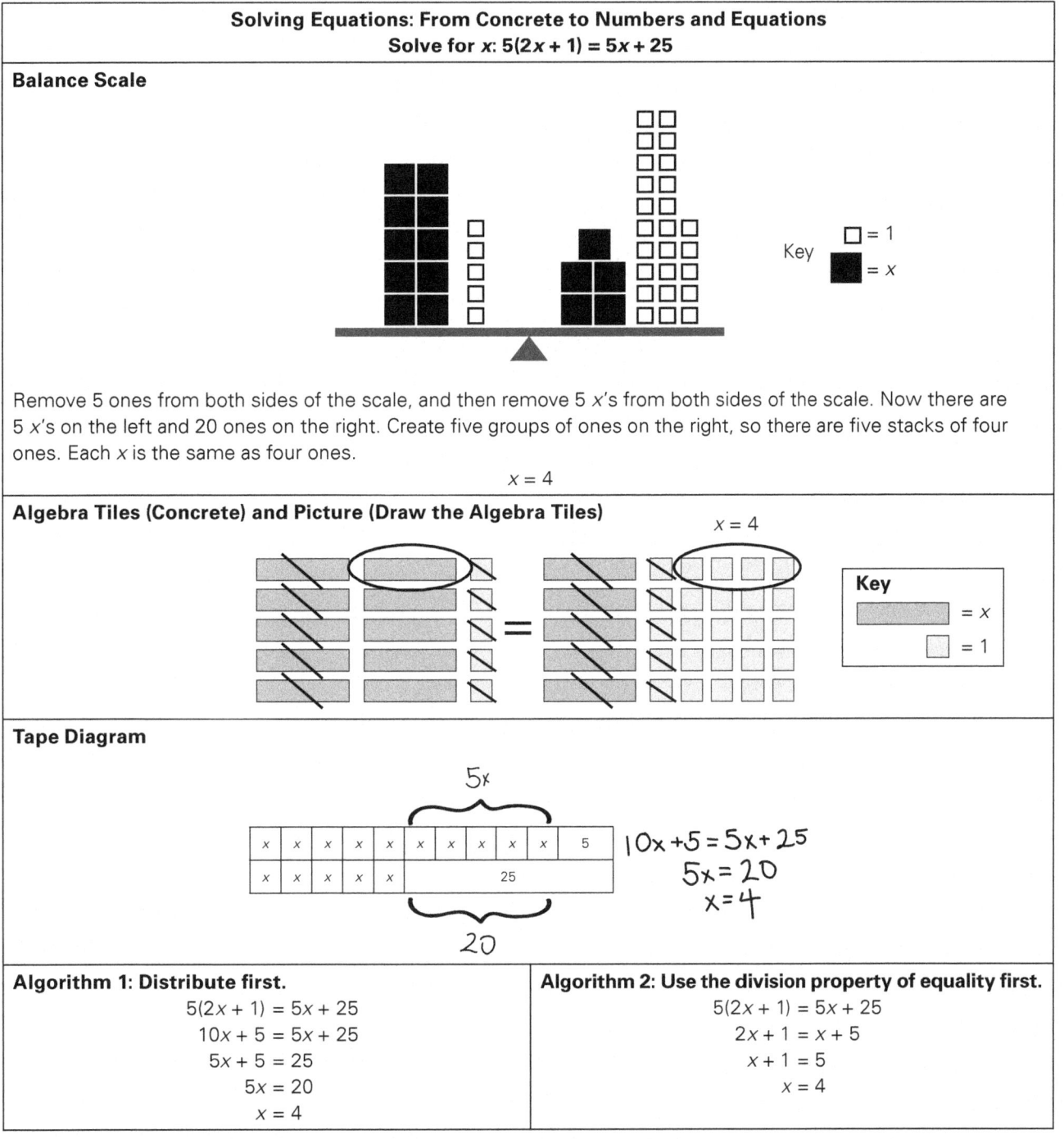

**Figure 7.4: Progression of strategies for solving equations.**

slope, and the *b* represents the *y*-intercept. Students plot a point for the *y*-intercept and determine a second point using slope triangles from the *y*-intercept.

High school students use knowledge of reflections, translations, rotations, and dilations to transform parent functions. Students might transform a function concretely using pipe cleaners on graph paper before moving to sketching technology-created graphs and eventually fluently graphing the transformation of a function by hand using the structure of terms in the function.

Procedural fluency includes flexible and efficient thinking—students who are developing an understanding of models, pictures, and abstract equations have more strategies to choose from when solving problems.

Discuss as a teacher team and with vertical grade- or course-level teams how students develop procedural fluency with the concepts taught in each grade. Strategies used in previous grades are often revisited in Tier 2 interventions.

### Existing Pictorial Models

Students may struggle at times to create their own picture representations for numbers and tasks in accurate and organized ways. Consider how your team can have students manipulate existing pictorial models to support them in drawing their own models. Students using a pictorial model to solve a task have a visual that represents values, allowing them to move more quickly to the operating phase of solving the task using numbers and symbols.

Figures 7.5 and 7.6 show examples of using existing pictorial models to make sense of the four operations and the distributive property.

Giving students drawings to work with during a task supports visualization and clarifies how to use drawing strategies. When asking students to describe their strategies orally, pictorial models provide a way to explain how to add, take away, or consider multiple groups.

### Strategy Connections and Comparisons

An important part of developing procedural fluency is the teacher's role in facilitating conversations that guide students to make connections between strategies. You and your team might do mathematics together to anticipate the strategies students will use when solving a task or analyze existing student work to identify strategies students are currently using toward procedural fluency. Discussing strategies can lead your team to ask, "Which strategies are most concrete? Which strategies are most efficient?" (Refer to figures 7.1–7.4, pages 100–103, for examples of different strategies students might use to make connections or compare.)

You support student learning during classroom problem-solving activities by circulating through the classroom, providing feedback, and identifying student work to share with the whole class. Plan how to engage students in dialogue as they compare and make connections between strategies. As a facilitator, share more concrete and less efficient strategies first, then make connections to pictorial models and representations before moving to more abstract pathways using numbers and symbols (Smith & Stein, 2018).

Figure 7.7 shows examples of guiding questions you can use to facilitate student thinking about comparisons and connections between strategies in small-group and whole-group learning settings.

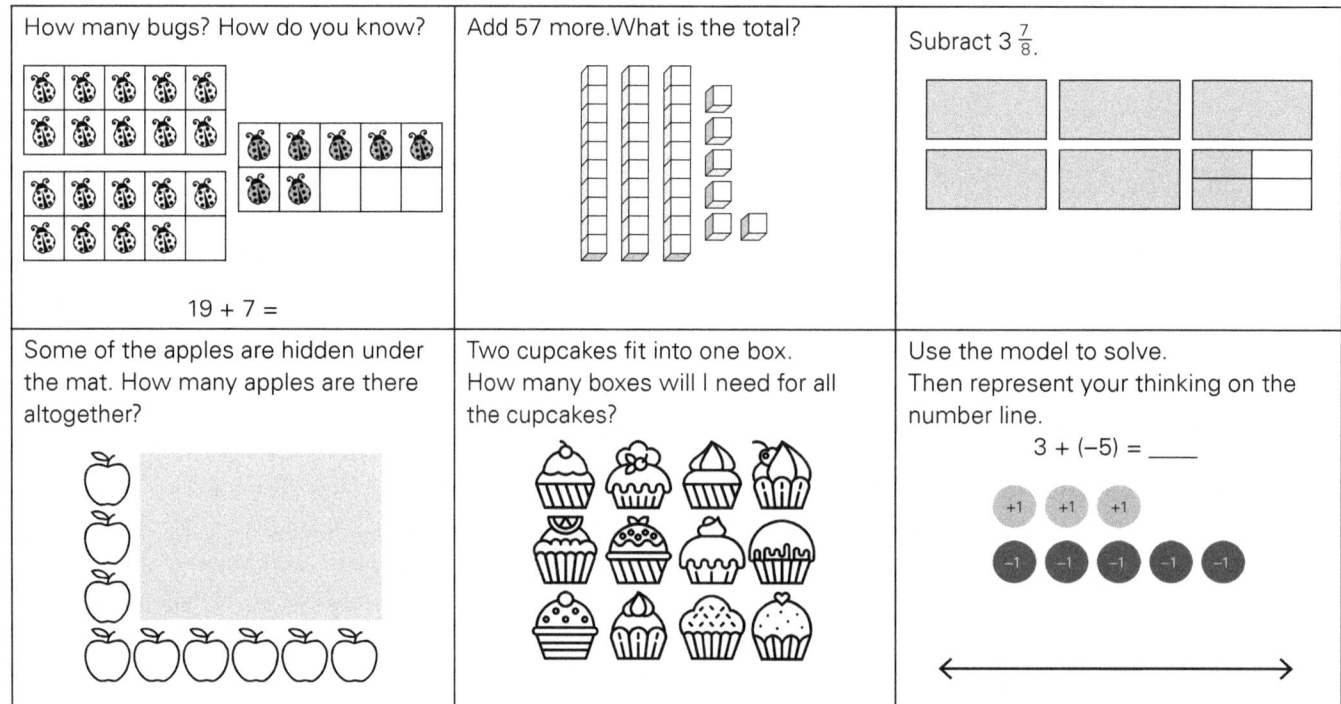

**Figure 7.5:** Elementary example of using existing pictorial models for the four operations.

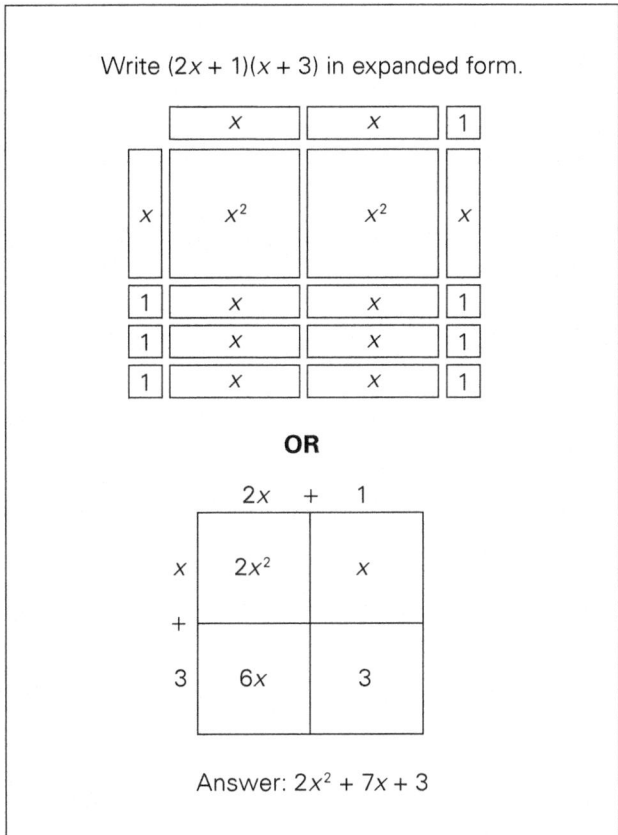

**Figure 7.6: Secondary example of using existing pictorial models to distribute.**

| Guiding Questions |
|---|
| How is your strategy similar to or different from [other strategy]? |
| What connection is there between your approach to solving the problem and the strategy shared? |
| Can you solve the same problem using fewer steps? |
| Which strategies make sense to you? Why? |
| Who can restate the strategy in their own words? |

**Figure 7.7: Sample guiding questions to facilitate connecting and comparing strategies.**

*Visit **go.SolutionTree.com/MathematicsatWork** for a free reproducible version of this figure.*

Recording student strategies on a class-created anchor chart allows you to return to discussions about choosing strategies as needed and can encourage students to use a more efficient strategy as part of giving real-time feedback. You support the process of moving students toward using more efficient strategies by allowing them to make connections and experience other solution pathways.

## Just Right Numbers

The Just Right Numbers strategy builds procedural fluency by asking students to show *why* a given answer is correct. It is developed from the Answers First strategy in *Engage in the Mathematical Practices: Strategies to Build Numeracy and Literacy With K–5 Learners* by Kit Norris and Sarah Schuhl (2016). Instead of students doing a mathematics problem to get an answer, an answer is provided, and students show the thinking and work that leads to that answer.

Create an anchor chart with several equations designed to be solved with procedural fluency, such as the one in figure 7.8, which can be used from second grade to fourth grade. Post the anchor chart in the classroom for a week or longer. Encourage students to choose two to four Just Right Numbers problems anytime during a mathematics lesson or small-group instruction and show why the given answer is correct. Change the number range and operations on the anchor chart to match your students' needs and your team's current procedural fluency goals. Providing answers and then asking students to prove the answers puts the learning emphasis on the work and minimizes incorrect practicing toward procedural fluency.

**Answers First!**
**Just Right Numbers**

| 67 − 29 = 38 | 500 − 249 = 251 | 425 − 193 = 232 |
|---|---|---|
| 376 + 92 = 468 | 91 − 64 = 27 | 710 − 163 = 547 |
| 900 − 598 = 302 | 874 − 358 = 516 | 83 + 47 = 130 |
| 52 − 19 = 33 | 408 + 166 = 574 | 619 − 344 = 275 |

You see the answer, but can your strategy prove it is correct?

**Figure 7.8: Just Right Numbers anchor chart example.**

In middle school, Just Right Numbers problems might involve developing procedural fluency with percentages, proportions, or solving equations, to name a few. The problems given on an anchor chart for percentages might look like 25% of 80 is 20. Students must show why 25% of 80 is 20. They might first explain using a ratio table, double number line, or base ten blocks. Eventually, students will use a proportion or percent equation to fluently show the percentage.

In middle school or high school, you and your team might want to use Just Right Numbers to help students develop procedural fluency with solving equations. You might give equations like $3x = 18$ with the answer $x = 6$ in sixth grade, $2x + 4.5 = 12.5$ with the answer $x = 4$ in seventh grade, and $2x^2 + 6x = -4$ with the answers $x = -1$ and $x = -2$ in high school. With several equations on a poster, students can pick some to show why the answer is true and how to generate it.

Differentiated practice does not need to involve hours of prep to be meaningful, nor is it accomplished by asking students to do pages of tedious tasks. Use and adjust routine practices, such as Just Right Numbers, as you teach different units with your team's selection of numbers in computations and equations.

### Over or Under Estimation

Estimation and reasoning about the accuracy of computational strategies are used in problem solving. Rounding and estimating are concepts that grow and support students developing procedural fluency. Students ask themselves, "What answer will make sense?" and after computing, ask, "Does this answer make sense?" The Over or Under strategy, shown in figure 7.9, engages students in estimating an answer to a computation without (or before) performing the computation. Figure 7.9 shows expressions from all grade bands.

The Over or Under strategy develops reasonableness without relying solely on rounding or performing a computation. Students generate their own approaches to estimating that make sense to them, with an emphasis on explaining their reasoning to support

procedural fluency. When given a task that lends itself to procedural fluency, students better understand what answer makes sense and check their thinking through the fluency strategies used to solve the task.

### My Favorite Mistake

A critical factor in students overcoming misconceptions is being aware of their errors as they solve problems and develop procedural fluency (see chapter 6, page 83, for more information about problem solving). Boaler (2016) shares:

> The recent neurological research on the brain and mistakes is hugely important for us as math teachers and parents, it tells us that making a mistake is a very good thing. When we make mistakes, our brains spark and grow. (p. 12)

Intentionally highlight common misconceptions and errors to engage learners in analyzing and fixing mistakes. Learning through mistakes is a shift in mindset that values productive struggle when students are learning. When developing procedural fluency, the tasks used for the My Favorite Mistake strategy focus on mathematics computations and number identification. Figure 7.10 shows an example task for the My Favorite Mistake strategy.

The My Favorite Mistake strategy uses teacher-created mistakes in student work, like figure 7.10, or is generated through student work in the classroom, as shown in figures 7.11 and 7.12 (page 108). Highlight a common error on an exit ticket and use the ticket as part of a whole-group discussion to find and fix the mistake before students make corrections on their own exit tickets. Model and think aloud during small-group instruction with the guiding question, "Can you find my mistake?"

A goal when developing procedural fluency is ensuring that students see making mistakes as a productive part of being a mathematician. As teachers raise

| Over or Under 100? | Over or Under 20? | Over or Under 1? | Over or Under 10? | Over or Under 5? |
|---|---|---|---|---|
| 46 + 73 | 203 ÷ 12 | $\frac{3}{8} + \frac{4}{9}$ | 48.25 ÷ 6.5 | $2 + \sqrt{6}$ |
| 58 + 29 | 325 ÷ 8 | $\frac{3}{5} + \frac{2}{3}$ | 35.12 ÷ 3.25 | $3\sqrt{3}$ |

**Figure 7.9: Over or Under strategy used for estimation.**

## Mathematics Task: Robert's Mistake

**Learning Goal:** My goal is to analyze a multiplication problem, find the mistake, and be able to correct the mistake by using a different strategy to solve the problem.

When Robert multiplied 36.23 and 100, he made an error.

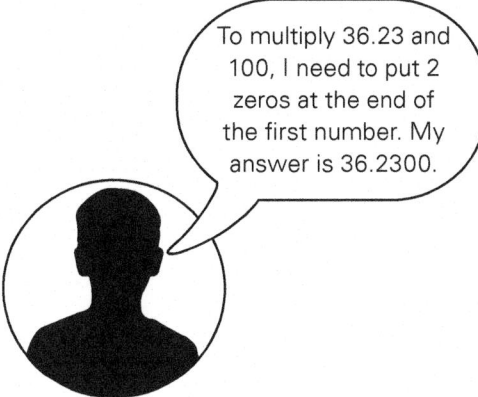

To multiply 36.23 and 100, I need to put 2 zeros at the end of the first number. My answer is 36.2300.

**Part A:** Use the strategy that works best for you to solve Robert's multiplication problem in the space provided.

**Part B:** In your own words, explain Robert's mistake and what he should do to get the correct answer.

**Before turning this in:**

Reread the learning goal.
- Did you meet the learning goal?
- Are you able to justify your response to a partner?

*Source: Adapted from Illustrative Mathematics, n.d.*

**Figure 7.10: Robert's mistake.**

awareness of errors, the likelihood of students making those same errors decreases, and their journey toward procedural fluency is accelerated.

### Card Sort

Another strategy for supporting procedural fluency through conceptual understanding is using a card sort. In a card sort, students are given a set of cards with each card showing a piece of information connected to the other cards. The card may show a mathematical term, pictorial representation, table, or graph, depending on the standard and grade level. Card sorting allows students to make sense of underlying mathematical structures and engage in meaningful discourse, and it fosters teamwork.

The goal of a card sort is for small groups of students to sort the cards into categories based on common characteristics or connections with an emphasis on number sense and understanding leading to procedural fluency. You and your team find or create a set of cards for students to sort, each with different information that can be categorized (for example, an equation, table, and graph; a percent, decimal, and fraction; or a numeral, ten frame, and picture of objects). You can use this activity in both Tier 1 and Tier 2 (and Tier 3 if the sort is developing fluency for something learned in a much earlier grade).

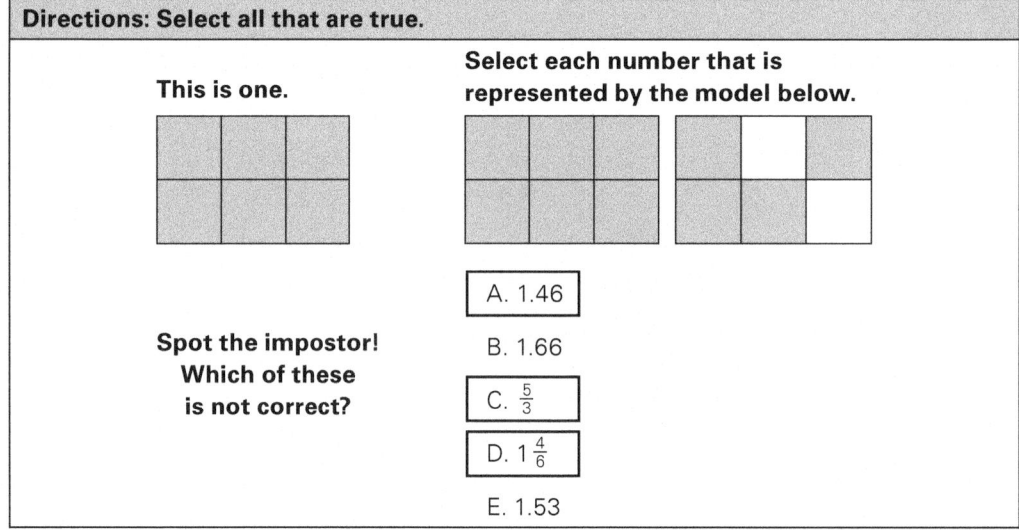

## My Favorite Mistake

**Directions:** A wrong answer has been circled.
What is partially correct about the choice? What is the mistake?
How would you fix the response? Justify your thinking!

**Directions: Select all that are true.**

**This is one.**

**Select each number that is represented by the model below.**

**Spot the impostor! Which of these is not correct?**

- A. 1.46
- B. 1.66
- C. $\frac{5}{3}$
- D. $1\frac{4}{6}$
- E. 1.53

## Overcome the Mistakes

**Directions:** Now it is your turn to learn from the mistake and use your understanding to solve similar problems.

Place a black dot on the number line that represents the decimal: $0.\overline{6}$. Write an equivalent fraction below the black dot.

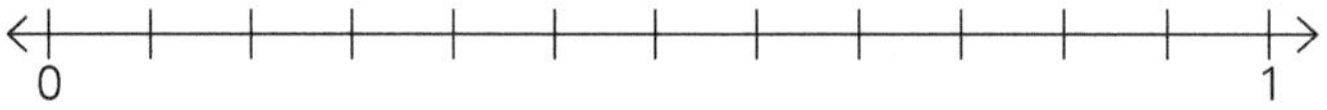

**Figure 7.11: My Favorite Mistake fraction and decimal example.**

**Directions:** The following shows student work for question 2 on the exit ticket. There is a mistake in the work. Identify the mistake and correct the solution.

Solve $3x^2 + 10x - 8 = 0$ using factoring.

$(3x + 2)(x - 4) = 0$

$3x + 2 = 0$ and $x - 4 = 0$

$x = -\frac{2}{3}$ and $x = 4$

1. Look at the answers on your exit ticket. Correct any errors below.

2. Solve using factoring. Show your work.

$2x^2 + 5x = 12$ $\qquad$ $5x^3 - 45x = 0$

**Figure 7.12: My Favorite Mistake solving quadratics example.**

Figure 7.13 (page 110) shows an example of a card sort using cards related to multiplication fact fluency. There are six cards for each multiplication fact that would be cut and separated for the sort. When used in Tier 1, give students a full set of multiplication by 5 facts, as shown in figure 7.13.

Ask students to quiz their partner by showing them a card without the multiplication fact and asking their partner to state the fact. Alternately, give students a full set of cards and have students match cards with the same numerical answer. When students match the cards, they explain to each other how the cards are equivalent. Another way to use the cards is to give each student a card, and then ask students to move around the room, finding everyone who has the equivalent numerical answer. During Tier 2, students use the cards to reinforce procedural fluency with partner activities.

Figure 7.14 (page 111) shows possible mathematical concepts for creating card sorts with your team to use in Tier 1 and Tier 2.

Card sorting in groups provides opportunities for students to make sense of multiple representations and promotes meaningful peer-to-peer discourse, which is explored more in chapter 9 (page 129). Use the same sets of cards multiple times in a variety of ways (games or different sorting instructions) with students as they develop procedural fluency.

## Conclusion

For students to achieve procedural fluency and apply it across various problem-solving settings, they must be exposed to concepts using concrete and visual models to make a stronger connection to abstract representations. Students develop an understanding of concepts and apply that learning to move toward procedural fluency when learning rigorous mathematics, meaning they eventually have efficient and accurate methods for solving problems.

Students in Tier 1 need exposure to different strategies, and time to understand which strategies work best in different situations, when developing procedural fluency. Identify key strategies aligned with the CRA framework for consistency across your grade or course. Work vertically with colleagues in your school (previous and future grade levels or courses) to develop procedural fluency using common strategies as students move from one grade or course to the next. Remember, procedural fluency is not a goal in mathematics—it is a way to efficiently and accurately apply an algorithm to solve a task. But understanding the mathematics behind any algorithm is critical so students have a tool kit of strategies to use. The strategies in this chapter can also be used for Tier 3 learning experiences for students still developing fact fluency or procedural fluency from years prior.

Students not yet demonstrating procedural fluency with mathematics facts or algorithms can learn procedural fluency through the intentional strategies and models your team uses in daily lessons and Tier 2 learning experiences. Identify first the desired procedural fluency and then how you will collectively use models and strategies concretely, representationally, and abstractly to develop that fluency.

## Questions to Consider for Next Steps

As a collaborative team, use the following questions to reflect on your current practices and determine any next steps for building number sense.

1. In what ways are you currently engaging learners using the CRA framework (concrete, representational, abstract)?

2. How are students getting opportunities to make connections between their strategies and those used by their peers when solving tasks?

3. How are you encouraging students to think flexibly and try more than one solution pathway to get accurate answers?

4. How are you visually representing varied strategies and models using class-created anchor charts?

5. In what ways are students estimating answers and evaluating their own accuracy?

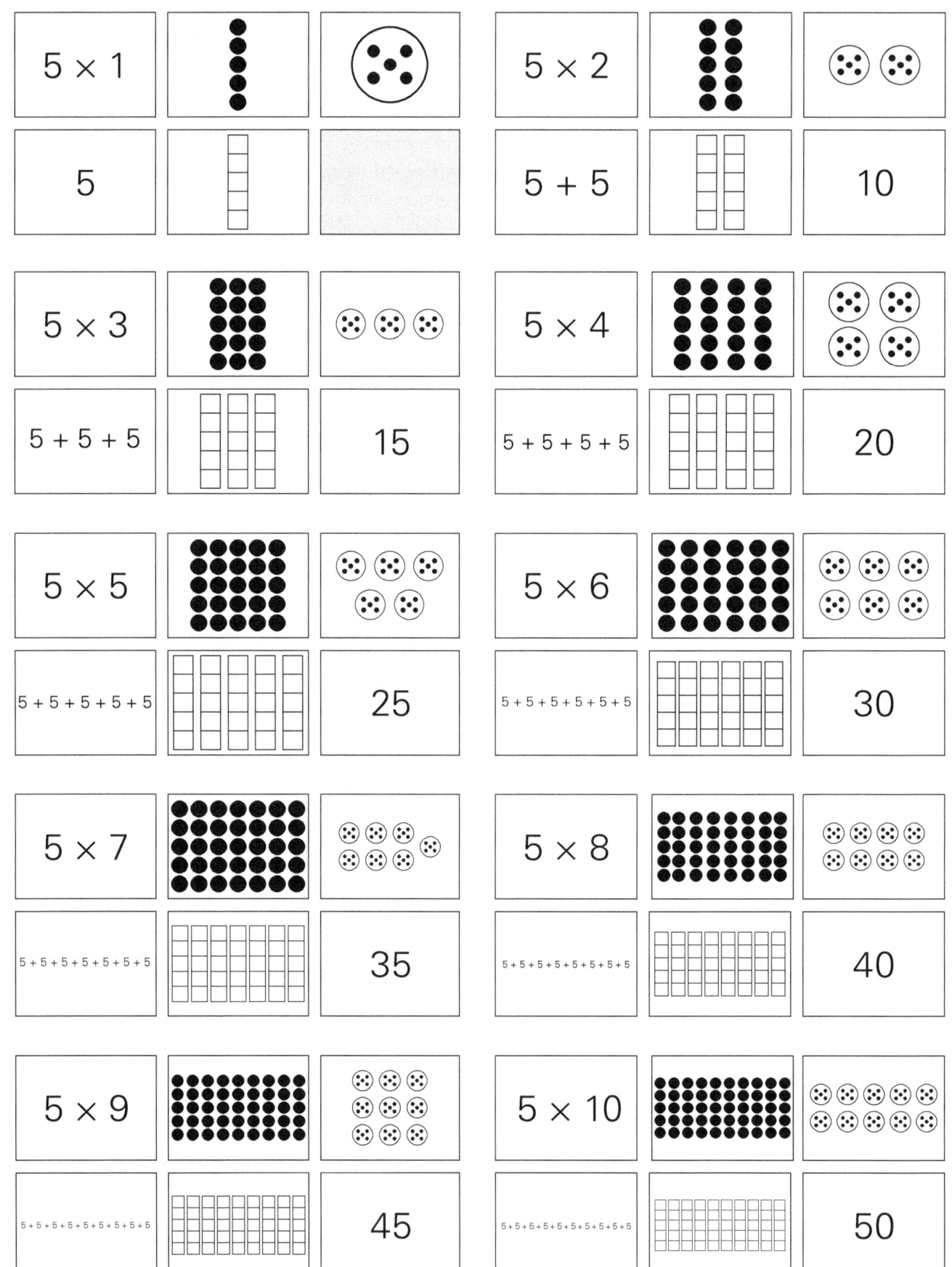

**Figure 7.13: Multiplication by 5 card sort.**

Visit **go.SolutionTree.com/MathematicsatWork** for a free reproducible version of this figure and other reproducible card sorts for facts for 3, 4, 6, 7, 8, and 9.

| Grade Band | Mathematical Concepts for Card Sorts |
|---|---|
| Grades PreK–K | • Object sort by shape, color, and size<br>• Number representations (subitizing) |
| Grades 1–2 | • Word problems, number sentences, and representation<br>• Addition and subtraction (with or without regrouping)<br>• Number sentences that are true or false<br>• Greater than or less than with numbers and representations |
| Grades 3–5 | • Quadrilaterals<br>• Comparing fractions<br>• Ordering fractions |
| Grades 6–8 | • Equivalent equations, solutions, and application problems<br>• Steps to solving equations<br>• Irrational square roots (number line approximation and square root)<br>• Exponent rules<br>• Measures of center |
| High School | • Functions (connecting the mathematical language, function, table, and graph)<br>• Types of angles and lines<br>• Proofs<br>• Solving equations |

**Figure 7.14: Examples of mathematical concepts by grade band for card sorts.**

*Visit **go.SolutionTree.com/MathematicsatWork** for a free reproducible version of this figure.*

# PART 3

# Student Engagement

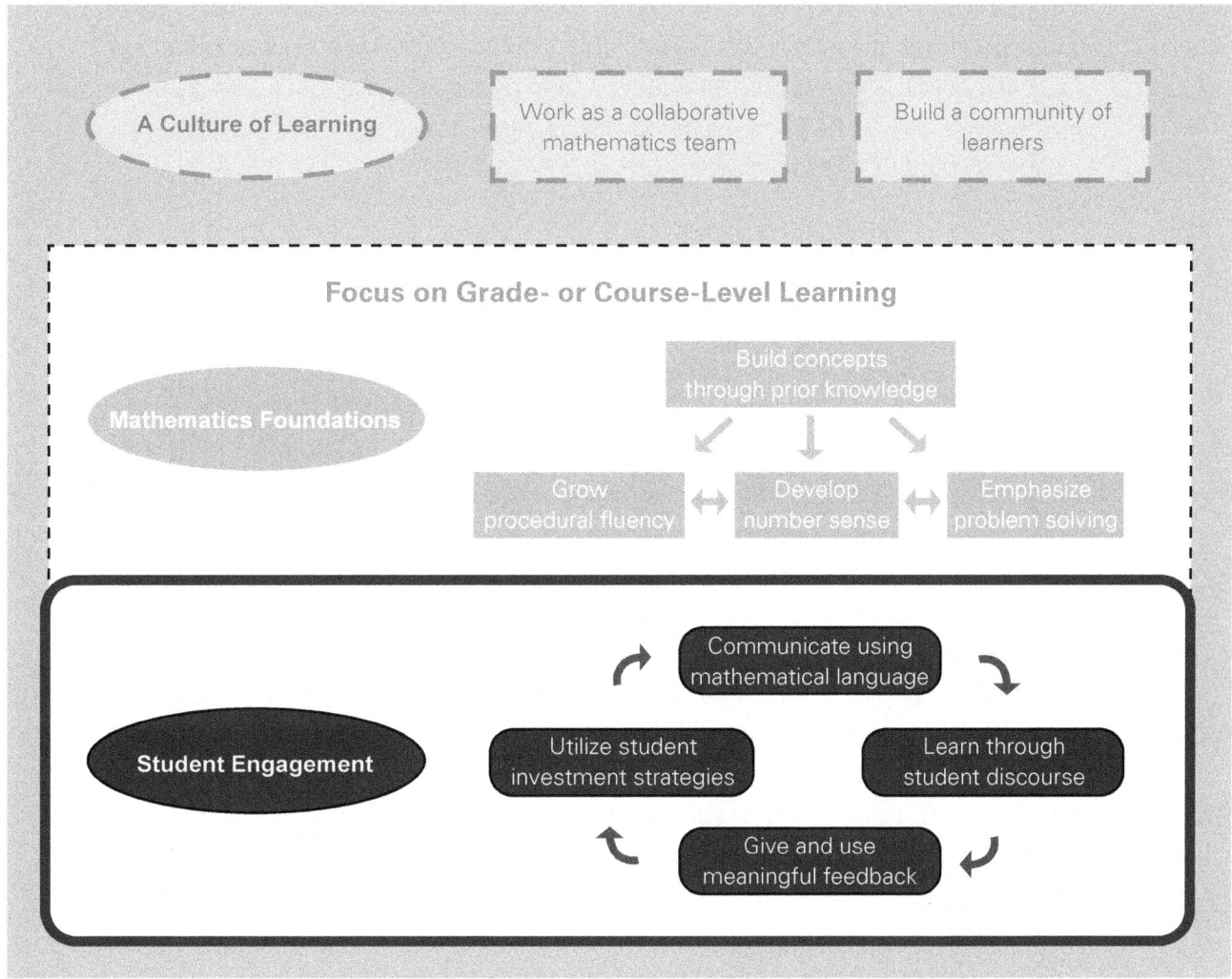

> At one K–5 school where Mona was coaching, Mona walked through classrooms and noticed the teachers using informal language and mnemonic devices to introduce mathematics problem-solving concepts. Interestingly, each grade level used a different mnemonic, but in all cases, each letter in the mnemonic stood for a problem-solving action. For example, in second grade, the teachers used *CUBES* as a strategy for word problems. In third grade, teachers used the mnemonic *SWEEP,* and fifth-grade teachers used *SOLVE.*
>
> | | | |
> |---|---|---|
> | C = Circle<br>U = Underline<br>B = Box<br>E = Evaluate and draw<br>S = Solve and check | S = Symbols<br>W = Words or labels<br>E = Equations<br>E = Explain<br>P = Pictures | S = Study the problem.<br>O = Organize the facts.<br>L = Line up the plan.<br>V = Verify the plan with computation.<br>E = Examine the answer. |
>
> Mona wondered, "Could the mnemonics be causing confusion for students?" The mnemonics changed each year, and each mnemonic used the letter E, which stood for different actions. Students seemed to be confused why problem solving changed based on the change in mnemonics and why the letter *E,* for example, also changed in meaning. Additionally, these problem-solving steps caused students to focus more on what they needed to write down for the mnemonic than the problem itself and their solution pathway. How were students making connections to learning the actual mathematics during problem solving?
>
> During classroom observations, when students were talking about their problem-solving pathways in small groups, Mona heard students say, "I moved that over here," "That number does this when I move it over," or "I added two, and it gave me the answer." As she captured the comments from students, she was not sure how well they were learning mathematics due to their vague explanations when communicating ideas. Students need to start with informal language; however, Mona observed a missed opportunity during the lesson when the teacher consolidated the learning and did not formalize the mathematical language students needed to describe their solution pathways.

CHAPTER 8

# Communicate Using Mathematical Language

> I've always enjoyed mathematics. It is the most precise and concise way of expressing an idea.
>
> —N. R. Narayana Murthy

Mathematics is its own language, and it is important for students and teachers to be as precise as possible. Instead of "I moved that over here," the student might have said, "I moved each digit of the number one place value to the left when I multiplied by 10" (see story on page 114). The second statement is clear, shows how the student is thinking, and allows other students to listen, learn, and give feedback.

In *Mathematics Instruction and Tasks in a PLC at Work, Second Edition* (Toncheff et al., 2024), mathematical language during instruction is described as follows:

> [Mathematical language is] language students use to make sense of and communicate the mathematics they are learning. Students will listen, read, write, and speak while engaging in solving mathematical tasks. Mathematical language includes knowing the meaning of and using words, symbols, notations, and abbreviations, which are important when students are learning essential standards. (p. 33)

Students who engage in mathematical discussions communicate with their peers and teacher through listening, speaking, writing, and reading (Garrett, Ritchie, & Phillips, 2022). Most importantly, students initiate and respond to mathematical ideas with one another. As the saying goes, the person who is doing the talking is the one doing the learning.

The U.S. Department of Education recommends teaching clear and concise mathematical language for effective Tier 1 instruction and Tier 2 interventions (Fuchs et al., 2021). It recommends supporting the use of mathematical language through the following:

- Routinely teach mathematical vocabulary to build students' understanding of the mathematics they are learning.
- Use clear, concise, and correct mathematical language throughout lessons to reinforce students' understanding of important mathematical vocabulary words.
- Support students in using mathematically precise language during their verbal and written explanations of their problem solving. (Fuchs et al., 2021, pp. 11–20)

Students learn when they communicate the mathematics during both Tier 1 and Tier 2 instruction. To accomplish this, students need time and space every day to learn and apply mathematical language in class.

## The *Why* and the *What* of Communicating Using Mathematical Language

Students strengthen their ability to communicate in class with peers, evaluate their progress on mathematical tasks, analyze their own thinking, construct an argument, and justify their reasoning when they learn and use mathematical language (vocabulary, symbols, and notations). Students understand and use mathematical language most effectively when they practice previously learned and new mathematical language daily (Toncheff et al., 2024).

Using precise mathematical language grows student learning in the following ways.

- Expands students' mathematics vocabulary and builds capacity to define or learn new terms
- Supports students' critical thinking about their ideas and the ideas of their peers
- Empowers students to clearly communicate and ask questions of their peers when they are solving problems
- Helps students refine their solution pathways and thoughts
- Assists students with making connections from one mathematics concept to another

Being mathematically literate means reading, speaking, listening, and writing mathematics vocabulary, numbers, symbols, and notations correctly to organize learning and mathematical reasoning. For example, do students know to read the symbol < as *less than*? In preK and kindergarten, do students know to read = as *the same as* or *equal* to develop an understanding of equality? Do students know what *tenths place* means?

The English language presents challenges that also affect mathematical language. Norris and Schuhl (2016) share some examples of how mathematical language and everyday words can cause confusion:

> Consider the multiple ways that some common words such as face, feet, and power are used. In everyday life, face and feet can refer to our own body parts or those of an animal. Mathematically speaking, face refers to a polygon making up one flat surface of a three-dimensional shape, and feet represents a standard unit of measure. Think of the word power, and we generally think of strength or amount of electricity. Power in a mathematical context refers to an exponent. Homophones can also indicate different meanings. For example, sum and some can cause confusion. (p. 152)

The English language can be very confusing, even to native speakers, and many schools have students working to become bilingual or multilingual once they learn English. The National Center for Education Statistics (2023) finds that 10.3 percent of U.S. public school students identify as English learners—about five million students. Learning mathematical language and developing a deeper understanding of the content, added to the complexity of English language acquisition, presents an additional challenge for many students.

Too often, teachers see this unique challenge as a roadblock to every student learning. Learning content in a new language is difficult and can become an excuse for why students are not learning. Consider, instead, how you and your team will emphasize mathematical language during lessons and continue to view multilingual students "as students who possess knowledge, strengths, and resources (i.e., asset-based rather than deficit-based lens)" (NCSM & TODOS, 2021, p. 1).

As you design high-quality mathematics learning experiences during Tier 1 and Tier 2 instruction as a team, discuss the mathematical language students need to learn to make sense of the content. Create plans for how students will effectively communicate their thinking using identified mathematical language. Too often, mathematical discourse is limited to answer-seeking conversations, tricks, mnemonics, or imprecise language, which limits students' ability to learn rigorous mathematics, including conceptual understanding and procedural fluency (explored more in chapter 7, page 95). Learning mathematical language, no matter how difficult the words or abstract the symbols, allows students to assign meaning to concepts and skills they can connect to prior, current, and future learning.

Students need equitable and consistent experiences with mathematical language across your team and vertically through the grades or courses in your school to be able to communicate with one another within any grouping in your classroom, across your team, and from year to year. Consider how you and your team create common expectations around the use of mathematical language. Table 8.1 shows how your team can support the development of strong mathematical language.

Mathematical language and discourse are addressed in two recommended mathematics teaching practices in NCTM's (2014) *Principles to Actions: Ensuring Mathematical Success for All* and two of the Standards for Mathematical Practice from the Common Core State Standards showing the habits of mind students need to develop when learning mathematics (NGA & CCSSO, 2010). Your state or province may have different mathematics standards, but every region includes habits of mind related to mathematical

Table 8.1: Communicate Using Mathematical Language in Tier 1 and Tier 2

| Communicate Using Mathematical Language in Tier 1 and Tier 2 ||
|---|---|
| <ul><li>Identify the mathematical language (words and notations) for each unit of study and lesson.</li><li>Discuss when and how to teach the vocabulary and symbols during the lesson (preteach, just in time, or formalize at the end of the lesson) to help students understand.</li><li>Use precise mathematical language during both Tier 1 and Tier 2 instruction.</li><li>Encourage using precise mathematical language in whole-group and small-group discourse.</li><li>Use students' informal mathematical language to help them make connections to formal mathematical language.</li><li>Encourage students to speak and write to ensure they have a clear understanding of mathematical language.</li></ul> ||
| **Communicate Using Mathematical Language in Tier 1** | **Communicate Using Mathematical Language in Tier 2** |
| <ul><li>Use mathematical language routines and graphic organizers to support language development and use.</li><li>Create word walls and graphic organizers during core instruction.</li></ul> | <ul><li>Determine how to re-engage students in making sense of the mathematical language learned during the unit.</li><li>Reference word walls or other graphic organizers used during Tier 1 to support Tier 2 instruction.</li></ul> |

language. The teaching practices describe the actions of effective mathematics teachers, and the Standards for Mathematical Practice describe the actions of mathematically proficient students. Table 8.2 shows how teachers and students are expected to use mathematical language for learning.

Students must precisely communicate to learn from one another through meaningful mathematical discourse and analysis of one another's work, as shown in table 8.2. Teachers support students' communication and their critiques of each other's work by using evidence of student thinking during both Tier 1 and Tier 2 instruction. To position students as owners of their mathematical knowledge, your team models and reinforces how to effectively communicate mathematics using clear and precise language. Additionally, teach students the routines of engaging in mathematical discourse (see chapter 9, page 129, for more information about mathematical discourse).

The first collaborative team action is to agree on the precision of the language you use to support student learning. The *Mathematics Unit Planning in a PLC at*

Table 8.2: Comparison of Teacher and Student Actions

| **Mathematics Teaching Practices**<br>*Effective mathematics teaching will:* | **Standards for Mathematical Practice**<br>*Mathematically proficient students will:* |
|---|---|
| **Facilitate meaningful mathematical discourse.**<ul><li>Help students share, listen, honor, and critique each other's ideas.</li><li>Help students consider and discuss each other's thinking.</li><li>Strategically sequence and use student responses to highlight mathematical ideas and language.</li><li>Model the use of questions and language expected during classroom instruction.</li></ul> | **Attend to precision.**<ul><li>Communicate precisely using appropriate terminology.</li><li>Specify units of measurement and provide accurate labels on graphs.</li><li>Express numerical answers with an appropriate degree of precision.</li><li>Write and say the correct symbols and notations used in making sense of a task or solving a problem.</li><li>Provide carefully formulated explanations.</li></ul> |
| **Elicit and use evidence of student thinking.**<ul><li>Identify strategies or representations that are important evidence of student understanding.</li><li>Make just-in-time decisions based on observations, student responses to questions, and written work.</li><li>Use questions or prompts that probe, scaffold, or extend students' understanding.</li><li>Strategically share student work from concrete to abstract or use another schema for students to make connections between their strategies and learn from one another.</li></ul> | **Construct viable arguments and critique the reasoning of others.**<ul><li>Make conjectures and use counterexamples to build a logical progression of statements to support ideas.</li><li>Use definitions and previously established results.</li><li>Listen to or read others' arguments.</li><li>Ask other students probing questions.</li><li>Write and orally communicate a solution pathway used to solve a task.</li></ul> |

*Visit* **go.SolutionTree.com/MathematicsatWork** *for a free reproducible version of this table.*

*Work* series provides resources and supports to identify the mathematical language students need to learn and communicate within each unit. Using imprecise vocabulary or relying on tips and tricks that do not promote conceptual mathematical understanding leads to confusion. Mathematics language includes vocabulary that is similar to words students use every day.

Review table 8.3 and consider words or phrases that may cause confusion for students. With your team, identify possible words or phrases that need to be revised for an upcoming instructional unit.

The list of words and phrases in table 8.3 is not exhaustive. However, it includes examples for your team to discuss when identifying how precise the mathematics language must be to support student learning and develop mathematical literacy.

## The *How* of Communicating Using Mathematical Language

In addition to being mindful of the mathematical language your team uses, consider the mathematical language routines used in your classrooms during Tier 1 and Tier 2 instruction. According to education researcher Jeff Zwiers and colleagues (2017), a mathematical language routine is:

> . . . a structured but adaptable format for amplifying, assessing, and developing students' language. . . . These routines can be adapted and incorporated across lessons in each unit to fit the mathematical work wherever there are productive opportunities to support students in using and improving their English and disciplinary language. (p. 9)

Table 8.3: Words or Phrases to Support Mathematics Understanding

| Words or Phrases That Cause Confusion | Words or Phrases That Support Mathematics Understanding | Mathematics Context or Rationale |
|---|---|---|
| Borrow or carry | Regroup | Add or subtract. |
| Plug in a number | Substitute a number | Replace a number for a variable in an expression or equation. |
| Reduce | Simplify | Rewrite a fraction using an equivalent fraction that does not have a common factor for the numerator and denominator. |
| Timesed, as in *I timesed 5 and 5*, or times, as in *5 times 5* | Multiplied, as in *I multiplied 5 and 5*, or groups of, as in *five groups of 5* | Use as an introduction to multiplication. |
| Read 4.85 as *four point eighty-five* or *four point eight five* | Read 4.85 as *four and eighty-five hundredths.* | Use *and* for the decimal point and use appropriate place value when reading or saying a decimal value. |
| Read 2 + 2 = 4 as *two plus two makes four* | Read 2 + 2 = 4 as *two plus two equals four* or *two plus two is the same as four.* | Read = as a sign showing the relationship between two sides of an equation. *Makes* suggests = is an action or operation. |
| Improper fraction | Fractions greater than one | Reference mixed numbers and fractions simply as *fractions*. |
| Goes into, as in *5 goes into 8 once*, when describing the division process | *How many groups of 5 are there in 8?* when describing the division process. | Multiplication is the inverse of division and *groups of* supports this connection. In later grades, 20 ÷ 5 would be read as *20 divided by 5*. |
| Evenly, as in *5 divides evenly into 20* | *20 divided by 5 is a whole number,* or *20 divided by 5 does not have a remainder.* | Divide two numbers without a remainder. |
| Read $\frac{1}{5}$ as *one over five.* | Read $\frac{1}{5}$ as *one-fifth.* | Read a fraction as a single number. |

This chapter's strategies are both instructional strategies and mathematical language routines used to foster communication and build student understanding. Consider how your team can implement these strategies and routines to grow student learning and develop precise mathematical language in Tier 1 and Tier 2 instruction.

## Word Walls

A word wall is a powerful strategy to help students clearly understand the mathematical language expectations during each lesson. Before each instruction unit, your teacher team articulates the vocabulary and notations students will be learning and communicating. As you introduce new mathematical language, post the words on chart paper, a bulletin board, or a whiteboard. As words from the unit appear during instruction, students write the words' definitions and give an example or draw a picture in their notes or mathematics journal to make sense of it.

Reference the word wall routinely in class to build consistency with this strategy. Ask students to use the word wall when communicating in class and have them include a chant read of the new mathematical language. A *chant read* is when the whole class repeats the word or phrase together several times to practice reading and speaking the word or phrase correctly. For example, the symbol < is on the word wall, and the example might be 3 < 5. Encourage students to have a partner discussion—partner A reads "three is less than five" to partner B, and then partner B reads it back to partner A.

You can also use a word wall for a class scavenger hunt. Give students a list of shape names (for example, quadrilateral, parallelogram, rhombus, trapezoid, rectangle, and square), and then ask them to find the shapes in the classroom or around the school. When learning how to classify shapes, ask students to name the figures given on a word wall and explain that some special shapes have more than one name. For example, a square is also considered a rhombus because it has four congruent sides, a rectangle because it has four right angles, a parallelogram because its opposite sides are parallel, and a quadrilateral because it has four sides. A right triangle might also be an isosceles triangle. Provide a sentence frame to help students describe each figure, such as: This is a _____ because _____.

Figure 8.1 shows a quadrilateral corner word wall.

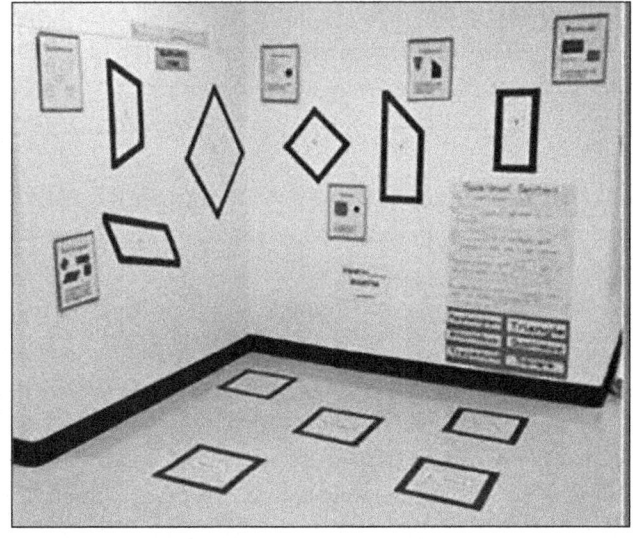

*Source: © 2013 by Jennifer Deinhart. Used with permission.*

**Figure 8.1: Quadrilateral corner word wall.**

You and your team can create word walls in your high school classrooms using the terms and mathematics notations for parent functions. Add the terms and any notations for new functions to the word wall. Students draw the parent function in their mathematics journal, and, after they have learned some or all the terms, they look at a variety of parent functions and sort the graphs using the word wall. Figure 8.2 shows an example high school word wall with parent functions. You might also add words for the key features of each type of function as students learn the functions.

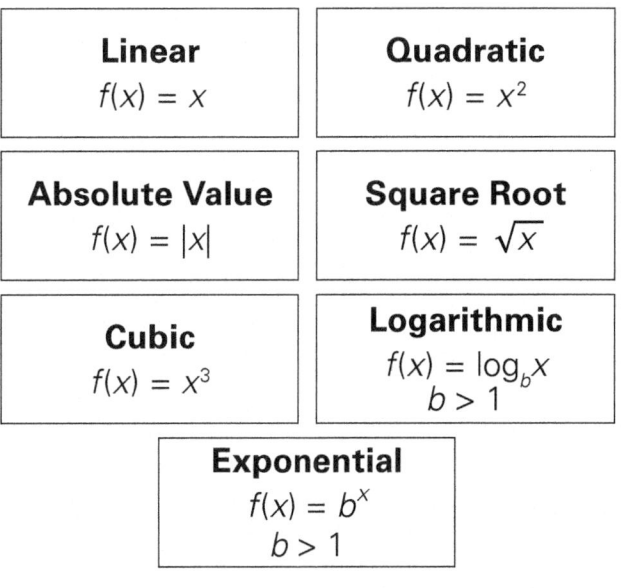

**Figure 8.2: High school word wall example showing parent functions.**

Word walls provide the language and notations needed when students engage in other mathematical language routines or use graphic organizers.

### Which One Doesn't Belong?

The Which One Doesn't Belong? strategy is often attributed to mathematics educator Christopher Danielson (2019). Danielson developed this strategy to engage students in mathematical thinking and discussion. Which One Doesn't Belong? involves presenting students with images of four different mathematical objects, figures, numbers, or expressions that have similarities and differences. Ask students to identify one image that does not belong and explain why. There is rationale for any one of the four images not belonging, so students are challenged to continue identifying an item and explaining why it doesn't belong.

The best part of this routine is there are no right or wrong answers; therefore, your collaborative team learns a lot about students' current knowledge and mathematical language through their communication. In mathematics, the images used in this strategy are usually numbers, shapes, or graphs. Students use precise mathematical language to explain their answers.

Consider figure 8.3 from Pam Wilson (n.d.; https://wodb.ca), which shows four numbers in a chart. Ask students why one (or all) of the numbers does not belong and to justify their answers.

|  |  |
|---|---|
| 9 | 16 |
| 25 | 43 |

**Figure 8.3: Which One Doesn't Belong? example.**

During a lesson, one group of students might state that 9 doesn't belong because it is the only single-digit number. Another student might say 9 doesn't belong because it is the only number whose digits do not sum to seven. Several students might think 43 doesn't belong because it is the only number that is not a perfect square or is the only prime number. Another student might say 25 doesn't belong because it is a multiple of 5 or is the only number that can be represented with a single coin (money). Finally, students might suggest 16 doesn't belong because it is the only multiple of 4 or is the only even number.

After a quick sharing, you have learned that students understand factors and multiples, prime and composite numbers, perfect squares, and even and odd numbers. If a student uses informal language in an explanation, you know the specific academic language to reinforce. Figure 8.4 shares additional examples of the Which One Doesn't Belong? strategy for your team to discuss.

By intentionally selecting the images used for Which One Doesn't Belong?, you learn from students what they already know about some content and the current precision level of their mathematical language.

### Would You Rather?

The Would You Rather? strategy originated from author Justin Heimberg's (2007) book series and game of the same name (Heimberg & Gomberg, n.d.), and it has since been adapted by John Steven for mathematics (Would You Rather Math, n.d.). In this activity, give students two situations or two images and have them choose which one of the pair they would rather engage with to answer a specific question and justify their reasoning with mathematics. The prompt typically pits two choices against one another, as shown in figure 8.5.

For this activity, the first step is to identify the prompt students will explore and the mathematics they might use to justify their choice. Be sure to clarify your expectations with students to encourage communication. Ask students to think about the prompt for a short amount of time, and then ask them to share their justification with a partner or student group. Remember, the most important element of this activity is to have students justify their reasoning and practice using precise mathematical language.

After students think independently about their answer and then discuss their thoughts with their partner or group, you might randomly select several pairs or groups to share their thinking with the class. It is not about right or wrong answers—it's about listening to students' thought processes. Use this strategy to develop the precise language students need to justify their reasoning.

**Directions:** Look at each example and discuss the mathematical language your team would expect to uncover if you used the example with students.

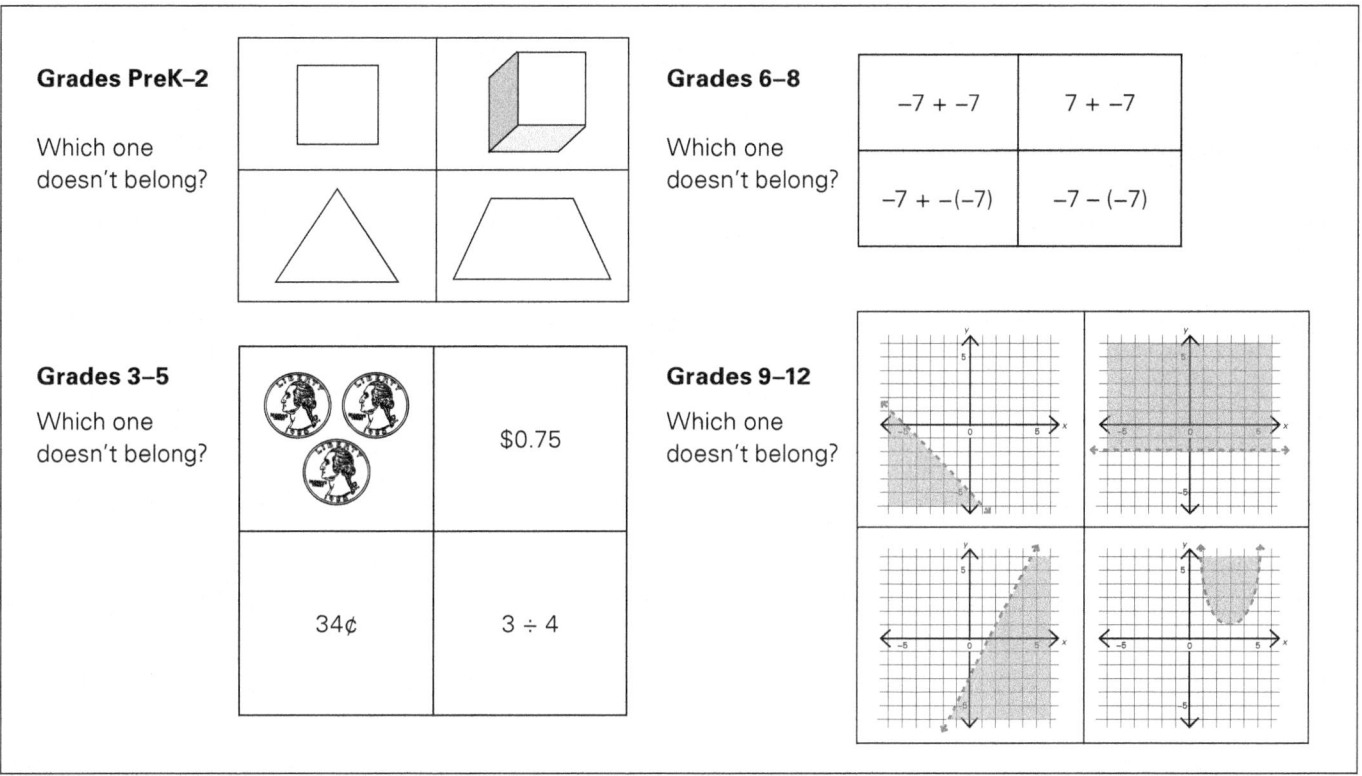

**Figure 8.4: Which One Doesn't Belong? examples.**

**Directions:** Read each prompt and discuss how it will reveal student understanding and use of mathematical language when students justify their answers.

| Grade Band | Would You Rather? Prompts |
|---|---|
| Grades PreK–K | • Would you rather have one more or one fewer cookie than your friend?<br>• Would you rather count to 100 by ones or skip count by tens? |
| Grades 1–2 | • Would you rather sit at a table where two children share five brownies or sit at a table where four children share seven brownies?<br>• Would you rather count *forward* or count *backward* to solve the following problems?<br>$\quad 4 + 6 = $ _____ $\quad 5 + $ _____ $ = 9 \quad 19 - $ _____ $ = 7$<br>$\quad 8 - 3 = $ _____ $\quad $ _____ $ + 57 = 73 \quad 41 - $ _____ $ = 22$ |
| Grades 3–5 | • Assume you're really hungry! Would you rather eat one-third of one-sixth of a pizza or one-half of one-eighth of a pizza?<br>• Would you rather have 435 minutes or 7 hours of play time per week? |
| Grades 6–8 | • Would you rather drink hot chocolate with two parts water and three parts cocoa mix or drink hot chocolate with three parts water and four parts cocoa mix?<br>• Would you rather buy a $50 shirt on sale for 20% off the original price, or a $50 shirt that was discounted 10% off the original price and then put on sale for an additional 10% off? |
| Grades 9–12 | • Would you rather put $10 in a bank and have it triple every week for 6 weeks or put $10 in a bank account and have it quadruple each week for 3 weeks?<br>• Would you rather solve $x^2 - 2x - 15 = 0$ using factoring or the quadratic formula? |

**Figure 8.5: Would You Rather? prompts.**

*Visit **go.SolutionTree.com/MathematicsatWork** for a free reproducible version of this figure.*

### Share-Trade Protocol and Stronger and Clearer Each Time

Share-Trade is an instructional routine you can use when giving students open-ended questions and you want them to refine their mathematical language over time. For example, in first grade, you might ask students to come up with as many ways as possible to represent one hundred. In fifth grade, you might ask students to add multidigit decimal numbers using any strategy. Or in high school, you might ask students to solve a linear or quadratic function. Ask students to respond to each prompt individually and then share their thinking with a partner. This continues as students move around the room and gain more insight into how to respond.

Educators Kerri Wingert and Abby Rhinehart (2016) describe the Share-Trade protocol as follows.

1. Students receive an open-ended question to answer.
2. Each student writes their individual thoughts about the question.
3. Students then move around the room to find a partner, taking their response with them.
4. Students must high five each other to form the partnership.
5. Each student shares their response and trades papers with their partner.
6. Each student is now responsible for sharing their partner's ideas, even if they do not agree with those ideas. This is not the time for a critique.
7. Students form partnerships three or four times and repeat sharing their thinking and their peers' thinking.
8. Students return to their seats and write a final solution.

The Share-Trade strategy is similar to the mathematical language activity Stronger and Clearer Each Time, created by Zwiers and colleagues (2017). Stronger and Clearer Each Time also provides an opportunity for students to revise and refine both their ideas and their verbal and written output. The Stronger and Clearer Each Time activity provides a purpose for student conversation by using discussion-worthy and iteration-worthy prompts.

For example, in second grade, you might offer students the prompt: *Imagine you have 30 candies. How many ways can you share them among your friends? Draw pictures or write number sentences to show your thinking.* In seventh grade, you might offer students the prompt: *Decide which lemonade is stronger—the first glass has 200 ml of lemon juice and 300 ml of water, and the second glass has 100 ml of lemon juice and 200 ml of water.* Start by giving students a rich and meaningful prompt that offers multiple strategies or ways of thinking to share. The following are directions for Stronger and Clearer Each Time.

1. **First draft response (two to three minutes):** Students think and write individually about a thought-provoking question or prompt.
2. **Structured pair meetings (one to two minutes for each meeting):** Students meet with two to three different partners. During each meeting, students take turns being the speaker (to share their ideas and writing), and the listener (to ask the speaker clarifying questions and give feedback).
3. **Final draft response (two to three minutes):** Students write a final draft. Final drafts are revisions of original work. They show evidence of incorporating or addressing new ideas, examples, and reasoning about mathematical concepts learned in their structured pair meetings. Final drafts are clearer, showing evidence of refinement in language and precision. Students compare their first and second drafts after they finish (Zwiers et. al, 2017).

The main idea for Share-Trade and Stronger and Clearer Each Time is to have students think and write individually about a question, use a structured pairing strategy allowing for multiple opportunities to refine their responses through conversation, and then finally revise their original written response. Subsequent conversations and additional drafts show evidence of incorporating their new ideas and mathematical language (including symbols and notations). The new drafts also show evidence of refinement in students' "precision, communication, expression, examples, and reasoning about mathematical concepts" (New Teacher Project, 2021).

## Mathematics Journaling

Students use mathematics journals during Tier 1 instruction and then reference their notes and original thinking during Tier 2 intervention. A mathematics journal provides opportunities for students to practice written communication and boost their mathematical confidence because they practice writing their thinking before sharing with peers. Research shows that students who practice using academic language boost their use of mathematical language and grow their mathematical understanding (Graham, Kiuhara, & MacKay, 2020; Kostos & Shin, 2010).

Mathematics journals may look different at the primary level. Students might create journals by stapling their work together from tasks completed, or they might receive a sentence frame to complete, such as *Today in math, I learned _____*. Students in the primary grades might also orally answer the prompt at the end of the lesson.

Mathematics journaling means more than just simply taking notes during a lesson. In *Building Thinking Classrooms in Mathematics, Grades K–12: 14 Teaching Practices for Enhancing Learning*, mathematics education professor Peter Liljedahl (2021) observes that, for too many students, the primary purpose of note taking is writing something down after the teacher says or writes it. Liljedahl (2021) finds that students are less engaged in mathematical thinking during the notetaking portion of the lesson when they are just asked to copy definitions or steps. Additionally, students copying notes are often unable to keep up with the teacher because they are still writing while the teacher continued instruction.

To ensure students are engaged in thinking, Liljedahl (2021) suggests students write down notes that are important to them at the end of the lesson, which he calls "notes to your forgetful self" (p. 193). Ask students at the end of a lesson in Tier 1 or following a Tier 2 session to record in their mathematics journals any new vocabulary, big ideas or concepts, procedures or steps, or examples that will jog their memory when they are working on independent practice. Figure 8.6 shows a graphic organizer that can be used in a mathematics journal.

| Today I learned . . . | Vocabulary I want to remember . . . |
|---|---|
| Notes to my forgetful self . . . | Examples . . . |

*Source: Adapted from Liljedahl, 2021.*

**Figure 8.6: Graphic organizer for meaningful notes.**

*Visit **go.SolutionTree.com/MathematicsatWork** for a free reproducible version of this figure.*

Visit the vibrant community of teachers on the Building Thinking Classrooms Facebook group (www.facebook.com/groups/buildingthinkingclassrooms) for additional examples of notetaking graphic organizers.

Students' graphic organizer notes and mathematics journal entries can sometimes reveal unplanned learning and insights. There are several types of mathematics journal entries that include reflection, problem solving, or content-specific prompts (McAnelly, 2021). When you and your team begin to identify, create, or revise writing prompts, be sure students are engaged in "mathematical writing" that calls "for students to reason mathematically" (Casa et al., 2016, p. 3). Mathematics education professor Tutita M. Casa and colleagues (2016) list five approaches to identifying, creating, and writing prompts that maximize mathematical thinking:

1. Promote students' solution paths.
2. Go beyond asking students to simply "explain."
3. Prompt students to share their reasoning.
4. Have students consider the validity of a given solution.
5. Have students debate the validity of two given solutions. (p. 540)

Figure 8.7 shows sample mathematics journal prompts to explore while thinking about how to use mathematics journals in your classroom.

When planning journal prompts, consider how students will include mathematical language they are learning (such as vocabulary, symbols, and notations) in their responses. Writing prompts that ask students to explain their understanding of a concept provides opportunities for your team to evaluate students' depth and level of mathematical understanding to better inform an instructional response in Tier 1 and Tier 2 instruction.

### Consensus Boards

As students are learning new content or mathematical language, your team can use large whiteboards to focus students on collaborating, communicating, and coming to an agreement on their mathematical thinking. The instructions in figure 8.8 explain how to use consensus boards for a group of four students during instruction.

Use this activity when you end a unit and review during Tier 1 instruction before the end-of-unit assessment. Modify the routine to make it more competitive when appropriate. You may also use the consensus board strategy as a formative assessment to provide quality feedback during an activity. Use consensus boards to obtain more information about student thinking during Tier 2 learning.

Coming to consensus is hard and sometimes emotional work. Remind students that coming to consensus is about debating ideas, not arguing with people. It is about critiquing the thinking, not the person. Consider scaffolding participation even further by holding class debriefs about how the consensus process went for different student teams.

### Mathematics Graphic Organizers

Graphic organizers are another tool to develop mathematical language. A graphic organizer is an instructional tool students use to organize and structure mathematical language to include words and symbols, concepts, and representations. The goal of a graphic organizer is to help students make sense of the mathematics they are learning through mathematical language and examples.

There are several graphic organizers you and your team can use to support the daily development of mathematical language—vocabulary and notations—during Tier 1 and Tier 2 instruction. You might use a KWL chart and have students write what they know about a concept before you teach it (such as fractions in fourth grade or linear functions in high school) and what they want to know. When the unit is complete, have them revisit the graphic organizer and write what they learned. A KWL chart usually has three columns titled: Know, Want to Know, and Learned. The graphic organizers in figure 8.9 (page 126) provide additional examples to support developing mathematical language.

| Type of Mathematics Journal Prompts | Example Prompts |
|---|---|
| Reflection Prompts | • Is mathematics your favorite subject? Why or why not?<br>• My best experience in mathematics was when . . .<br>• Describe how today's mathematics lesson supported your learning.<br>• Explain everything you know about _____. |
| Problem-Solving Prompts | • (Provide sample student work and then ask the following question.) Is the work correct? Explain why or why not.<br>• Write and solve a word problem whose solution involves . . .<br>• What model or strategy would you use to solve . . . ? |
| Content-Specific Prompts | • What do you know about the number 5?<br>• How do you know if two fractions are equivalent?<br>• How do you simplify . . . ?<br>• Explain the proof of the Pythagorean theorem.<br>• Teach a friend the key features of the graph of the function $y = x^2 + 10x + 9$. |

**Figure 8.7: Sample mathematics journal prompt.**

*Visit **go.SolutionTree.com/MathematicsatWork** for a free reproducible version of this figure.*

**Setup:** Each small group or table needs a large sheet of chart paper or whiteboard. One student on the team equally partitions the rectangle and leaves an oval in the middle that is the consensus center.

1. Each student claims one of the rectangular corners of the chart paper or whiteboard.
2. The teacher poses a mathematics question to solve.
3. All students get a predetermined amount of time to think and write their thoughts in their respective corners.
4. Students take turns explaining their ideas to each other (all students must share).
5. Students discuss what their consensus solution might be and write their final response in the middle. All student work, regardless of solution pathway, must support the consensus answer.

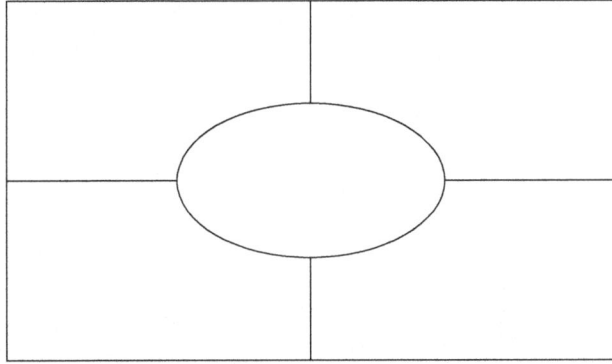

**Figure 8.8: Consensus board for student groups of four.**

Graphic organizers help students break apart the content and the mathematics language. They are also a great tool to support problem solving, which is explored in more detail in chapter 6 (page 83).

## Conclusion

Students need to develop their mathematical language to communicate precise thinking. That starts with you and your team determining what mathematical language students need to learn and communicate in each unit and modeling it during daily lessons. Work with your team to identify current strengths in student mathematical language and any routines, strategies, or activities you may want to incorporate into Tier 1 and Tier 2 instruction. Communicating by reading, writing, speaking, and listening allows students to make sense of mathematics and articulate what they are learning, as well as what they learn from others.

## Questions to Consider for Next Steps

As a collaborative team, use the following questions to reflect on your current practices and determine any next steps related to growing students' ability to communicate using mathematical language.

1. When does your team discuss the mathematics language that students must learn when teaching the essential standards and big ideas?
2. How does your team explicitly teach mathematics vocabulary?
3. What are some potential language barriers or tricky mathematics words that you need to teach?
4. How can you develop student understanding of mathematical language (both words and notations)?
5. What mathematical language activities, routines, or strategies can you add to your instructional tool kit?

| Graphic Organizers |
|---|
| **KIP**<br>*KIP* is a graphic organizer that uses a chart to document important vocabulary: key vocabulary, information about the vocabulary, and a picture drawing of the word. The five steps are the following.<br>1. Create a chart like the following.<br><br>      \| **Key** \| **Information** \| **Picture** \|<br>      \|---\|---\|---\|<br>      \| \| \| \|<br>      Example:<br><br>2. Students document the key vocabulary (K) in the chart. Each word will have its own chart.<br>3. Students write information (I) about the word—their own definition for the word.<br>4. Students draw a picture (P), if appropriate, for the word.<br>5. Students write or draw an example showing the word. |
| **Frayer Model**<br>This is a graphic organizer for vocabulary.<br>1. Have students write the word they are going to define in the middle of the graphic organizer.<br>2. In the upper-left corner, have students write the definition of the word (in their own words).<br>3. In the upper-right corner, have students write the facts or characteristics they know about the word.<br>4. In the lower-left corner, have students write or draw an example of the word.<br>5. In the lower-right corner, have students write or draw a nonexample of the word.<br>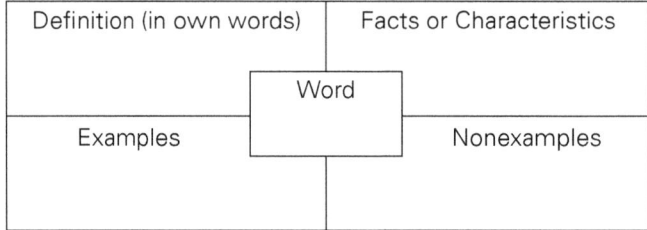 |
| **Foldable**<br>A foldable is a tool used to organize vocabulary for any unit. Most foldables require one sheet of paper that you fold or cut to creatively organize words, definitions, and examples. You can use foldables to integrate reading, writing, thinking, data organizing, researching, and other communication skills into an interdisciplinary mathematics curriculum (Zike, 2003). Visit the *Math Equals Love* blog (https://mathequalslove.net/category/inbs/foldables/page/12) for more information about using foldables.<br>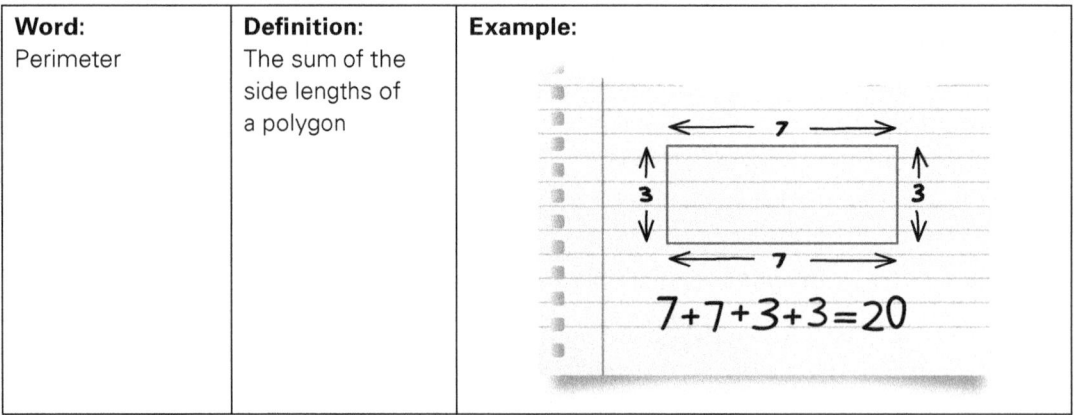 |

Source: *Dale & O'Rourke, 1986; Frayer, Fredrick, & Klausmeier, 1969; Zike, 2003.*

**Figure 8.9: Mathematics graphic organizer examples.**

*Visit **go.SolutionTree.com/MathematicsatWork** for a free reproducible version of this figure.*

> During a mathematics lesson, Mrs. Smith noticed the same students always raised their hands to respond when she called for answers. Question after question, the same small group of students volunteered to share their thoughts. Often, these were the students Mrs. Smith expected to know the correct responses. She was not confident the other students were learning like those with their hands raised. This was not the first time Mrs. Smith had seen this in her classroom. Any time she taught or reviewed in a whole-group setting, the same four or five students readily responded.
>
> Occasionally, Mrs. Smith would bypass the sea of usual hands in the air and ask for other volunteers, even going to great lengths to provide an anxiety-inducing amount of wait time for a response. She even attempted to use popsicle sticks with students' names written on them to randomize which students would respond. When Mrs. Smith was successful in soliciting responses from reluctant students, those students often resorted to random guesses, shrugged their shoulders in uncertainty, or simply responded, "I don't know."
>
> During a coaching session with Dr. Buck, Mrs. Smith inquired about what she might do to engage more of her students (specifically those who needed more support in mastering grade-level standards) during classroom instruction and in opportunities to learn with and from their classmates whose learning was advancing.

CHAPTER 9

# Grow Learning Through Student Discourse

*Effective mathematics teaching engages students in discourse to advance the mathematical learning of the whole class.*

—*National Council of Teachers of Mathematics*

It is all too common to have only a few students respond to questions posed to the entire class during instruction. In the story on page 128, Mrs. Smith was grappling with an important question: How can I ensure every student in my classroom is thinking about the questions being posed, learning from each other, and feeling confident enough to share a response with the class? Just expecting students to learn from one another through discourse is not enough. Students often need to be taught how to have effective conversations with one another.

As previously mentioned, the one doing the talking is the one doing the learning. When only a few students and the teacher are talking, as in Mrs. Smith's class, they are typically the ones doing most of the learning. Through meaningful discourse with one another, students verify their thinking and learn from their peers' feedback. Whether in a whole-group or a small-group setting, whether Tier 1 or Tier 2, it is a missed learning opportunity when students are not allowed or expected to share their thinking with each other.

## The *Why* and the *What* of Growing Learning Through Student Discourse

Discourse, or communication, among students is a way for them to take in others' thoughts, which is then a springboard for students to reflect on their own thinking (NCTM, 2014). As students listen to peers explain their thinking, they have the opportunity for their own thinking to be validated or challenged and to deepen their understanding or application of a concept. Likewise, as students are talking or sharing their own thoughts, they make their thinking clear to the listener (NCTM, 2014). Student discourse may begin with teacher-led prompts and questions, but it grows into student-generated conversations.

In *Adding it Up: Helping Children Learn Mathematics* (Kilpatrick et al., 2001), the National Research Council (NRC) concludes that student discourse provides an opportunity for students to (1) understand key mathematical ideas and (2) develop reasoning and problem-solving skills that contribute positively to a student's disposition toward mathematics. The NRC advocates for teachers to intentionally plan opportunities in daily lessons to grow mathematics learning through student discourse with these two focuses in mind. Once your team identifies a unit's key mathematical ideas and any connections to previous units, the challenge is to have students discover those connections through your selection of tasks and their work with one another. Students take more ownership of their mathematics learning through their discussions.

Along with teaching students grade- and course-level content standards during instruction, students also need to learn how to engage in mathematical processes (NCTM et al., 2021). Of the eight Standards for Mathematical Practice, which are the mathematical habits of mind or processes students must develop, the third practice states that students need to construct viable arguments and critique the reasoning of others (NGA & CCSSO, 2010). To convey their thinking and arguments to others, students may need to first use concrete objects or drawings and diagrams to clarify their thinking so they can share it more effectively (NGA & CCSSO, 2010).

When you structure consistent student discourse routines around open-ended mathematics questions, students have opportunities to give feedback to one another and themselves. Hattie's research shows an effect size of 0.82 for classroom discourse, which is equivalent to about a 29 percent gain in learning (Visible learning Meta$^X$, 2023). Hattie shares that student discourse has the potential to considerably accelerate student learning when it is student to student and student to teacher.

One of the six elements of high-quality lesson design in *Mathematics Instruction and Tasks in a PLC at Work, Second Edition* (Toncheff et al., 2024) is mathematical discourse routines. Teachers intentionally plan for student-to-student discourse throughout a lesson, whether engaging every student in answering questions posed to the whole group or encouraging students to interact with one another in small-group learning experiences. Not only does student discourse promote learning and student agency, but it also provides an opportunity to gather information and learn about student thinking during lessons. Toncheff and colleagues (2024) state the following:

> The manner in which you facilitate student discourse is essential in creating and supporting a classroom learning environment that values reasoning and sense making from the student's point of view. How you ask questions and implement discourse routines in your classroom has important implications for whether your instruction promotes a classroom culture that values students' voices, their experiences and ideas, and their cultures. (p. 58)

You and your collaborative team intentionally design learning experiences focused on students communicating with you and their peers in ways that ensure each student's voice is heard. Consider the following questions from authors Margaret S. Smith, Michael D. Steele and Mary Lynn Raith (2017) when reflecting on your use of classroom discourse:

- Are all students' ideas and questions heard, valued, and pursued in the mathematics classroom?
- What mathematical ideas does the class examine and discuss?
- Whose thinking does the teacher select for further inquiry, and whose thinking does the teacher disregard during small-group and whole-class discussion?
- Who in the classroom is positioned as competent?
- Whose ideas are featured and privileged? (p. 95)

The classroom culture during best first instruction (Tier 1) and intervention (Tier 2) impacts students' desires to talk and learn from one another (see chapter 2, page 27, for more information about building a community of learners). Student discourse is an instructional practice designed to grow learning in both Tier 1 and Tier 2 experiences and needs to be taught. Ideally, students will be curious about mathematics and ask the teacher and each other questions. Additionally, in both tiers, students will answer whole-group questions using think-pair-share or other strategies. A natural way to engage students in learning from one another through student discussions is with your choice of mathematics tasks. Does the mathematical reasoning in the task warrant a discussion? Remember that the language activities and routines used in the classroom impact the quality of student discourse (see chapter 8, page 115, for more information about mathematical language and student discourse).

Although student discourse is an instructional strategy to grow learning in both Tier 1 and Tier 2, there can be a few distinctions, as shown in table 9.1.

Students do not always understand their role in learning from each other in the classroom, especially if they have previously been taught to listen to the teacher, take notes, and practice steps when learning mathematics. Student-to-student interactions focused on learning from one another need to be taught. Even when first asked to discuss an answer to a problem, a student might say, "I got 7," and another might say, "I got 10," only to have both students sit quietly instead of discussing why they have different answers, waiting in silence until the teacher calls everyone back. Students know how to initiate the questions, What did

Table 9.1: Grow Learning Through Student Discourse in Tier 1 and Tier 2

| Grow Learning Through Student Discourse in Tier 1 and Tier 2 ||
|---|---|
| <ul><li>Design intentional opportunities for students to engage in productive dialogue with their peers throughout lessons.</li><li>Provide structures to support student-to-student discourse.</li><li>Create environments where all student voices are valued, heard, and appreciated.</li><li>Watch and listen to student-to-student discourse to inform instructional decisions.</li><li>Emphasize students' use of precise mathematical language.</li><li>Engage students in both whole- and small-group discourse.</li></ul> ||
| **Grow Learning Through Student Discourse in Tier 1** | **Grow Learning Through Student Discourse in Tier 2** |
| <ul><li>Ensure conversations focus on learning grade- or course-level mathematics standards and connect to prior knowledge, when needed.</li><li>Have students share and remind one another of the multiple representations and strategies that can be used to solve a task.</li><li>Have students eventually internalize and engage in mathematical conversations without the support of conversation structures.</li><li>Promote open dialogue related to the learning targets to encourage students to make connections.</li></ul> | <ul><li>Ensure that conversations focus on specific targets in essential grade- or course-level standards or the applicable prior knowledge targets, as determined by the targeted intervention.</li><li>Have students share insights about a single representation or strategy they are learning as part of the targeted intervention.</li><li>Ensure conversations are supported by structures, such as prompts or sentence frames, to focus on a specific aspect of mathematical learning.</li><li>Design specific conversations around students' areas of needed support related to the learning targets.</li></ul> |

you get? or How did you do it?, but in both instances, they tend to mean, Please give me the answer, so I can be done with the task. Meaningful student discourse focuses on students learning mathematics through engaging and purposeful peer interactions.

Consider the following group discussion between a group of three students while working on the third-grade mathematics problem: *During a fundraiser, 6 students sold a total of 48 cakes. How many cakes did each student sell if they all sold the same number of cakes?* (Refer to figure 3.5, page 51, for the task and the essential standard.)

**Student 1:** Hey, I am struggling to solve this task. I know that there are 48 cakes, and 6 students sold that many. Can you explain to me why we need to multiply these two numbers?

**Student 2:** I disagree because we need to find out how many each student sold. If we multiply the two numbers, the number of cakes gets larger, but they do not sell more than 48 cakes.

**Student 3:** I chose to create a tape diagram—see how the total is 48, and the 6 students represent each of the 6 boxes?

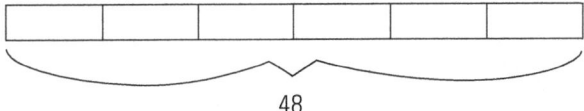

**Student 1:** That makes sense. So, when I look at your tape diagram, I think we need to divide to find out how many cakes each student sold.

**Student 3:** Do you think we could also multiply 6 times something to get 48?

The tasks your collaborative team chooses to implement in lessons, and the expected routines you teach, impact student learning through discourse. Discourse is often more meaningful when you are using high-level-cognitive-demand tasks that challenge students. See appendix B (page 179) for descriptions of high-level-cognitive-demand tasks.

# The *How* of Growing Learning Through Student Discourse

Student discourse is a valuable tool used to support, strengthen, and accelerate student learning. Student discourse begins with the tasks you choose as a teacher team—the tasks that give students something to talk about. Give students tasks with multiple entry points and that can be solved using a variety of strategies or solution pathways. Recall the definition of mathematical rigor in chapter 1 (page 15). If the task is practicing procedural fluency, students are more apt to complete their work individually, and any conversation is often forced. In cases of high-level-cognitive-demand tasks that focus on conceptual understanding and application, students more naturally talk together as they solve and make sense of the activity.

Too often, students have difficulty beginning and sustaining conversations, or even putting their thoughts and actions into words. The following example strategies are designed to help students share their thinking with one another and engage in meaningful conversations about the mathematics they are learning.

### Purposeful Questioning

Quality student-to-student conversations are prompted by the questions you and your colleagues pose throughout a lesson. Not all questions have the power to sustain meaningful, mathematically centered conversations. Through a meta-analysis of research studies, NCTM (2014, 2020) identifies "pose purposeful questions" as one of its eight Mathematics Teaching Practices that make mathematics learning accessible to and rigorous for all students (NCTM, 2014, p. 10). Further, NCTM (2014, 2020) synthesizes from their research that teachers generally ask four broad question types: "(1) gathering information, (2) probing thinking, (3) making the mathematics visible, and (4) encouraging reflection and justification" (NCTM, 2014, pp. 36–37).

- When gathering information, "students recall facts, definitions, or procedures" (NCTM, 2014, p. 36). A teacher might ask, "What is a rectangle?" or "What is the formula for the volume of a sphere?"

- When teachers are probing thinking, students explain their thoughts about strategies used to solve tasks. A teacher might ask, "Why did you use an area model to multiply?" or "Why did you choose to graph the two linear equations when solving the system?"

- Making the mathematics visible means students make connections between concepts and notice mathematical structures. A teacher might ask, "How does your equation match the situation in the word problem you are trying to solve?" or "How do the cubes show composing 10?"

- For the last question type, students are tasked to reflect on their learning and justify the validity of their work. A teacher might ask, "How do you prove the answer is 12?" or "How do you know rigid transformations preserve congruence?"

Use all four question types to stimulate discussion and discourse when working with students. NCTM (2014) states that it's not just the types of questions you ask, it's the pattern of your questioning that determines the reasoning required of students to answer. If you only use gathering information questions during instruction, students will be limited to lower-level mathematical thinking. However, using all the types of questioning, and routinely asking encouraging reflection and justification questions, engages students in high-level-cognitive-demand thinking.

Smith and Stein (2011) highlight three specific question types that allow students to think deeply about ideas: (1) exploring mathematical meanings and relationships, (2) probing, and (3) generating discussion. Table 9.2 shows sample questions for each of these three specific types.

Use the question types in table 9.2 during both Tier 1 and Tier 2 instruction. Questions that explore mathematical concepts, probe student thinking, and generate meaningful discussions can grow students' understanding of mathematics. Consider how you will use purposeful questions daily in future lessons. Incorporating specific questions into instruction based on lesson goals guides the focus of students' work and discussions.

Table 9.2: Sample Mathematics Questions to Deepen Student Thinking

| Question Type | Sample Questions |
|---|---|
| Exploring mathematical meanings and relationships | • How would you retell the problem in your own words?<br>• Would it help to create a representation, make a table or model, or draw a picture?<br>• What strategies have we learned previously that were useful in solving this problem?<br>• How does this concept relate to _____?<br>• Can you give me an example of _____?<br>• How might you prove _____?<br>• Do you notice a pattern? If so, can you explain it? |
| Probing | • Can you explain what you have done so far?<br>• What else is there to do?<br>• Why did you decide to _____?<br>• Can you think of another method that might work?<br>• What do you notice when _____? |
| Generating discussion | • What do you think about the solution pathway _____ shared?<br>• Do you agree? Why or why not?<br>• Does anyone have the same answer but a different way to explain it?<br>• How can you convince us that your solution makes sense?<br>• What do you notice when _____?<br>• Does anyone else have a more efficient strategy? |

*Source: Adapted from PBS LearningMedia, n.d.; Smith & Stein, 2011.*

*Visit **go.SolutionTree.com/MathematicsatWork** for a free reproducible version of this table.*

### Sentence Frames

Students may need support crafting the language to convey their thinking, methods, and strategies with their peers. Sentence frames provide students with language to structure their thoughts and focus their conversations. Figure 9.1 shows sentence frames from Norris and Schuhl (2016) that students can use both in their responses to teacher questions and to initiate conversations with peers across grade levels.

---
• I know because _____.
• Can you explain _____?
• I agree/disagree with you because _____.
• I understand this part, but I am confused by _____.
• This _____ (answer, step, or part) does not make sense because _____.
• I still have a question about _____.
• I think there is another way to _____.
• I wonder if _____.
---

*Source: Norris & Schuhl, 2016, p. 72.*

**Figure 9.1: Sample sentence frames to build discourse routines.**

*Visit **go.SolutionTree.com/MathematicsatWork** for a free reproducible version of this figure.*

Sentence frames can also be content specific. Figure 9.2 shows examples of content-specific sentence frames and frames that may contain more than one blank to fill in.

---
**Elementary Examples**
• My number is _____. It has _____ tens and _____ ones.
• My equation represents _____ groups, with _____ in each group.
• I can decompose the mixed number into _____ (place values).
• I know the two fractions are equivalent because _____.

**Secondary Examples**
• The area of a _____ (shape 1) is related to the area of _____ (shape 2) because _____.
• I know the slope of line _____ is steeper than line _____ because _____.
• The best measure of center for the data set is _____ because _____.
• The key features of the _____ (type of function) are _____.
---

**Figure 9.2: Sample content-specific sentence frames.**

*Visit **go.SolutionTree.com/MathematicsatWork** for a free reproducible version of this figure.*

You and your team can incorporate sentence frames into both Tier 1 and Tier 2 instruction in a variety of ways. One way to incorporate sentence frames is by posing a question to the class and having students use a specific sentence frame to respond to the question with partners or in small groups. For example, you might ask if the product of two numbers is always, sometimes, or never greater than the largest factor. Students discuss in their groups and answer with: *The product of two numbers is _____ greater than the largest factor because _____.*

Another option is to post several choices on an anchor chart or bulletin board—as shown in figure 9.3—and encourage students to choose an appropriate starter to initiate and support their conversations. Your teacher team might also make a bookmark with sentence frames for students to place in their textbooks or journals and use during discussions or tape sentence frames to each student's desk so they are easily accessible. As students practice using sentence frames, encourage them to create their own sentence frames as they see fit and share them with the class.

## CONVERSATION Starters

* I know _____ because . . .
* Can you explain _____?
* I agree/disagree with you because . . .
* This part does not make sense to me because . . .
* I feel most confident about _____.
* Another way to think about it is to _____.

*Source: © 2023 by Brian Buckhalter. Used with permission.*

**Figure 9.3: Sample anchor chart with sentence frames.**

For primary students, model the sentence frames with pictures or create talking sticks with verbal cues. Figure 9.4 shows examples of talking sticks.

**Figure 9.4: Example talking sticks.**

Ask students to turn and talk to their designated partner and share their thinking using talking sticks during carpet time. Begin by introducing the talking sticks and modeling with one student before the initial task. You can then begin a number talk. For example, students are exploring the number of the day, which is 14. Ask students to think about the number 14 and how many ways they can represent it. Then, ask partner A to share their idea with partner B. Partner B chooses one of the talking sticks to respond to partner A's thinking.

**Partner A:** I think I can represent 14 with 1 ten and 4 ones.

**Partner B [holding up the "I agree because" talking stick]:** I agree because 1 ten and 4 ones are the same as 14.

Set goals for student conversations when planning lessons, answering the questions, What do I want students to gain from talking to each other? and What mathematics do I want students to learn in their discussions? Also consider the grouping structures needed to promote meaningful student discourse.

### Flexible Grouping

Students benefit from various forms of grouping within the classroom to support student-to-student discourse, including random grouping, strength-based grouping, and grouping structured around students' readiness (NCSM, n.d.). However, these groupings should be able to be rearranged in terms of their size and makeup in a strategic manner to support learning based on students' needs (Van de Walle et al., 2019).

It is important to consider the outcomes for specific group structures when making decisions about grouping students to facilitate discussion. Random heterogeneous groupings during Tier 1 core instruction promote a classroom culture where students see each other as valuable learning resources. As one of his fourteen practices to building thinking classrooms, Liljedahl (2021) advocates using groups of three students and randomly assigning students to new groups every day. Students recognize that the groups are randomly assigned, which demonstrates to them that everyone can learn from any student in the classroom.

You might create groups using a deck of cards and ask students to draw a card when entering the classroom. They form groups with like cards. Or, you can use a computer program like Flippity (www.flippity.net) to input student names and randomly assign groups shown publicly on the board. You may also want to group students strategically, which is why the grouping is flexible.

During Tier 2 instruction, place students in small groups with a targeted learning focus based on data from common assessments. Student learning accelerates to grade or course level through their interactions with each other and intentional lesson design that addresses standards in a learning progression leading to the essential standard. At times, students may benefit from partner grouping for a task most suitable for two people, while other times they benefit from larger group configurations of four with specific roles for each person, such as when solving a high-level-cognitive-demand task (Toncheff et al., 2024).

For groupings to be successful, Van de Walle and colleagues (2019) recommend establishing both individual and shared accountability with students. *Individual accountability* means each member of the group will individually commit to understanding and explaining the processes, content, and product related to the task. *Shared accountability* means students work together to accomplish their common goal.

Consider how every student is engaged in student discourse while working in assigned groups and how every student is held accountable for learning. You may need to assign a role to each student on a team. Role examples include the following.

- **Recorder:** The recorder writes down the group's thoughts.
- **Questioner:** The questioner moves the conversation forward and clarifies group thoughts through questions.
- **Seeker:** The seeker asks the teacher for assistance and brings the information back to the group.
- **Manager:** The manager keeps track of time and maintains group focus on solving the task (Norris & Schuhl, 2016).

Regardless of role, any student may be asked to share their group's thoughts with others in the class, which helps with accountability. For example, you might ask students to show their work on poster paper for classmates to see during a gallery walk, with a student from each group standing with their poster as an expert to clarify the group's work or answer any questions.

You might also use the Three Stay, One Stray activity, in which a student from each group is assigned to take the group's work and move counterclockwise to the next group and explain the group's solution or thinking. Another idea is to use a team huddle. Have one student from each group come to the middle of the room to ask questions from other groups and then report back to their own group. Students who share their group's work are not assigned but are instead randomly selected after the group has worked on a task to hold all students accountable to learning. This grows student discourse throughout the task and while sharing solutions and group thinking.

To show that you encourage and expect student discourse in the classroom, arrange desks in close proximity for each group—rows will not work (see chapter 2, page 27, for ideas on desk arrangements for student teams). Frequently move students randomly in Tier 1 and use small strategic groups in Tier 2 interventions focused on specific learning targets.

### Jigsaw

The jigsaw strategy has the potential to considerably accelerate learning. In Hattie's research, jigsaw has an effect size of 1.20 (Visible Learning Meta$^X$, 2023). A jigsaw first arranges students into home groups. Each student in the home group is assigned a subtopic or problem. Students from each home group with the same subtopic or problem come together to form an expert group. Students discuss solution strategies within their expert group and respond to their portion of the task.

After conferring within their expert group and mastering their subtopic or problem, students report their findings to their home group and lead their peers in learning more about their subtopic or problem. Student discourse is naturally built into each of the groupings used in a jigsaw. Figure 9.5 shows how students rearrange themselves during a jigsaw activity.

Using the jigsaw strategy works well with high-level-cognitive-demand tasks with multiple parts or with a set of practice problems. For example, suppose second graders are learning the following standard:

> 2.NBT.B.7: Add and subtract within 1,000 using concrete models or drawings and strategies based on place value, properties of operations, and/or the relationship between addition and subtraction; relate the strategy to a written method. Understand that in adding or subtracting three-digit numbers, one adds or subtracts hundreds and hundreds, tens and tens, ones and ones; and sometimes it is necessary to compose or decompose tens or hundreds. (NGA & CCSSO, 2010)

The following is an example of the jigsaw strategy used to teach this standard and engage students in learning through discourse.

1. Give each home group a copy of the following questions.

    a. When should you regroup place values (for example, regrouping 100 into 10 tens)?

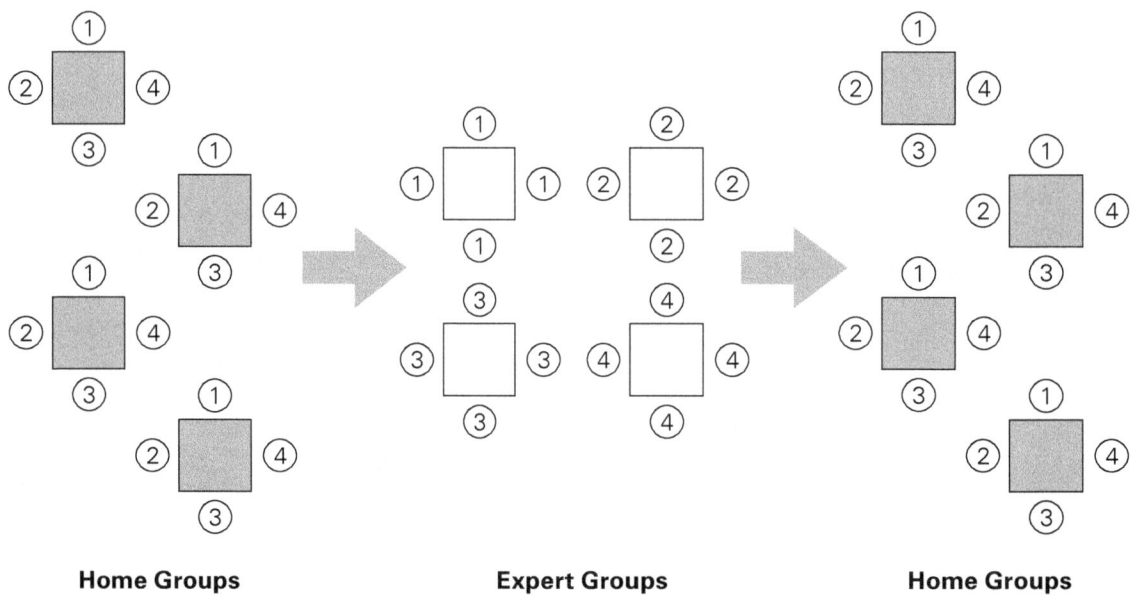

**Home Groups**      **Expert Groups**      **Home Groups**

**Figure 9.5: Student movement during a jigsaw activity.**

b. To subtract 432 – 176, will you need to regroup? Why or why not?

   c. Create pictures of base ten blocks to show how you would add 286 + 49.

   d. Create pictures of base ten blocks to show how you would subtract 286 – 49.

2. Number students one through four in their home groups, ask them to join the expert group with the same question number, work their problem with their expert group, and return to their home group to share the strategies and responses generated to answer their question.

You and your team might use a jigsaw for solving a system of equations in eighth grade or high school grades. Jigsaw also works for learning how different strategies are connected, as in the following example.

1. Pose a system of linear equation word problems and have home groups create the two equations for the system to use to solve the problem.

2. Give students a number one through four and have them go to their expert groups for: (1) solving by graphing, (2) solving using a table, (3) solving by substitution, and (4) solving by elimination.

3. Experts go back to their home groups to share how they solved the task. Students in the home group choose their favorite strategy and explain why using the sentence frame: *My favorite strategy to solve this problem is _____ because _____.*

You may want to use very specific sentence frames for both the home group and expert group student discussions (refer to figures 9.1–9.4, pages 133–134). Jigsaw works in both Tier 1 and Tier 2 to promote learning through student discourse.

## Conclusion

Orchestrating opportunities for students to engage in purposeful discourse with peers supports students' understanding of the mathematics they are learning. Providing students with questions or sentence frames offers guidance on how and what to discuss with their peers and structures discourse routines. Through consistent practice using sentence frames and other conversation guides, students will eventually generate and sustain their own conversations.

Along with providing support in facilitating their conversations, orchestrate opportunities for students to engage in discourse with peers through flexible groupings that can be random or based on students' needs. Utilizing strategies, such as jigsaw; gallery walks; or Three Stay, One Stray, helps students learn from one another through discourse and requires students to ask questions and explain their reasoning. Students must see each other as valuable learning resources in the classroom and then engage in student-generated, meaningful, and directed discourse to make sense of the mathematics they are learning.

## Questions to Consider for Next Steps

As a collaborative team, use the following questions to reflect on your current practices and determine any next steps related to student discourse.

1. How do you currently teach students to learn from one another using intentional student discourse?

2. How can students deepen their understanding of essential learning standards through discourse?

3. What vocabulary and mathematical language do you expect students to use while communicating their thinking, and how will you teach them to use it when engaged in conversations with each other?

4. How will you and your colleagues use grouping structures to grow student learning through discourse?

5. What are some of your favorite sentence frames that support student-to-student discourse?

> In Sarah's early career as a high school mathematics teacher, she dutifully taught as she had been taught—begin teaching a unit, give a quiz midway through the unit, teach the rest of the unit, and give an end-of-unit assessment. She spent hours grading quizzes and fixing mistakes or reasoning errors for students so they could learn from the feedback and be ready for the end-of-unit assessment. Too often, she found quizzes crumpled in the recycle bin or watched students glance at their scores and cram the quizzes into their backpacks. Students responded similarly to Sarah's feedback on end-of-unit assessments. She was left wondering why they did not seem to learn from her help in class or from her comments on their assessments.
>
> Later, Sarah attended a freshman basketball practice. She watched as students actively applied their learning on the court and received immediate feedback regarding their efforts from the coach and other players as they practiced in small groups. During a scrimmage, the players applied their learning and received more relevant and meaningful feedback related to their play as a team and as individuals. At the end of the week, the team was prepared for game day. And, after game day, they watched film of the game to learn from what they did right and any mistakes they made so they could continue strengthening practices to prepare for future games.

# CHAPTER 10

# Use Meaningful Feedback for Learning

> The key to learning is feedback. It is nearly impossible to learn anything without it.
>
> —Steven D. Levitt

In the story on page 138, Sarah provided feedback during lessons and assessments, but she did not ask students to take immediate action on her feedback. Unlike basketball scrimmages and games where feedback was seen as a valuable learning tool, classroom feedback was viewed as more judgmental than helpful. Sarah graded every activity and assessment, even when learning to proficiency wasn't yet expected.

To improve student learning, classroom feedback should be more like the feedback on the basketball court. In this chapter, we discuss ways for students to immediately act on feedback to build confidence and student agency, take ownership of their learning, and deepen their understanding of mathematics without worrying about a grade.

## The *Why* and the *What* of Using Meaningful Feedback for Learning

*Feedback* is information shared with students that identifies strengths in their mathematics reasoning and mistakes or errors to learn from. Effective feedback is critical to quality instruction in both Tier 1 and Tier 2. Feedback allows students to know and celebrate their growth in learning, and it also provides a way for teachers to learn about student thinking and respond accordingly through targeted instruction. Educational researcher Robert J. Marzano (2017) defines *feedback* as "the information loop between the teacher and the students that provides students with an awareness of what they should be learning and how they are doing" (p. 6). Learning happens through feedback and action.

The article "Inside the Black Box" by educational researchers Paul Black and Dylan Wiliam (1998) is foundational to the discussion of feedback. In the article, Black and Wiliam (1998) share the importance of quality feedback to students and teachers through frequent formative assessments: "We know of no other way of raising standards for which such a strong prima facie case can be made" (p. 148). The formative assessments Black and Wiliam (1998) refer to include the informal formative assessments done throughout daily lessons and the more formal common assessments your team uses to monitor student learning throughout a unit. The information gathered from daily lessons and common assessments creates opportunities to learn about and respond to student thinking and for students to then learn from their feedback.

When working to accelerate learning in mathematics, second-grade students, for example, need feedback on their ability to understand and apply strategies for addition and subtraction; seventh-grade students need feedback on their ability to understand and apply proportional reasoning; and high school students need feedback on their ability to understand and apply strategies for solving quadratic equations. Feedback to students about their current understanding alone will not necessarily help them know what to do next or see how close they are to learning a grade-level standard.

Professor and education researcher John Hattie's effect size for feedback showing students what they currently understand, what still needs to be understood, and how to move forward in the learning is 0.52. He also shares that feedback has the "potential to considerably accelerate learning" (Visible Learning Meta$^X$, 2023). Hattie (2023) emphasizes that feedback needs to be tied to a learning goal and success criteria to be understood and acted on by students (see chapter 11, page 153, for more information about student investment strategies). In other words, if the second-grade, seventh-grade, or high school student receiving feedback on their respective subjects recognizes the essential learning standard they are working toward and the success criteria (student skills) necessary for a full understanding, the feedback is more meaningful and actionable by students.

Additionally, Hattie's (2023) work emphasizes the need for student-to-student feedback. Such feedback occurs when students learn together in groups during lessons, solve tasks together, and formally analyze each other's work during class.

There are different types of feedback you and your team can give to students. Education professors John Hattie and Helen Timperley (2007) describe feedback as *personal-level feedback* (focused on the learner), *task-level feedback* (focused on the task), *process feedback* (focused on processes that can be replicated), and *self-regulation feedback* (feedback students give themselves to monitor, regulate, and direct their learning). See table 10.1 for examples of each feedback type.

Task-level feedback may not support learning if students do not understand a concept yet. Therefore, for intervention purposes, focus more on process-level and self-regulation feedback using success criteria and essential learning standards. Students' belief in their ability to do and understand mathematics grows when they monitor their own learning and determine next steps, leading to greater student self-efficacy and a positive mathematics identity.

Feedback is used well when it continues to grow student (and teacher) learning. Academic researcher Valerie J. Shute (2008) emphasizes the following factors for giving effective feedback.

- **Motive:** Does the learner need this feedback?
- **Opportunity:** Does the learner get feedback when it is most useful?
- **Means:** Does the learner have the willingness and capacity to learn from the feedback?

Students need frequent feedback in Tier 1 and Tier 2 mathematics learning experiences but may not need it when they are productively persevering through a task. Instead, consider each student and their needs in terms of the feedback. Education authors Cassandra Erkens, Tom Schimmer, and Nicole Dimich (2017) write: "When people are on the path to improvement, they are *on* the path. Students at the very early levels of the journey will benefit more from further instruction than feedback" (p. 44).

Table 10.1: Types of Feedback

| Type of Feedback | Description | Example |
| --- | --- | --- |
| Personal-Level Feedback | Feedback is about the student and may be vague. Personal-level feedback is not productive and implies fixed ability. | • Good job!<br>• Try again!<br>• You struggle with fractions. |
| Task-Level Feedback | Feedback is focused on a specific task. It is best used when students have some knowledge of the concept in the task. | • How do you see the dots in the image to find the total?<br>• Which part of the question have you answered? (Often used for two-step word problems.)<br>• What does the variable in your equation mean? |
| Process Feedback | Feedback is focused on a mathematical process that can be replicated in future tasks and continue to be improved. | • How might you keep track of regrouping ones to tens when adding?<br>• What is another strategy you might use to graph the linear function? |
| Self-Regulation Feedback | Feedback is generated by students who monitor their own learning and determine next steps to continue learning using their classwork and assessment data. | • What have you learned so far about multiplying two two-digit numbers?<br>• What are you still working to learn about finding the volume of pyramids and cones? |

*Source: Adapted from Hattie & Timperley, 2007.*

When done well, students develop a growth mindset using feedback. Feedback focused solely on effort without a connection to student work will, instead, create a fixed mindset (Dweck, 2015). For example, some students may think, "Even with effort, I can't do it," while others think, "With easy effort, I can do it." While effort is important, students grow their learning through feedback tied to their actual work so they can determine what they learned and what they still need to learn.

Schuhl and colleagues (2024) and Toncheff and colleagues (2024) refer to FAST feedback (fair, accurate, specific, and timely) when describing essential characteristics of formative feedback (Hattie, 2012, 2023; Hattie, Fisher, & Frey, 2017; Reeves, 2011, 2016).

- **Fair:** The feedback is focused on the student's work, not the student's characteristics.
- **Accurate:** The feedback correctly gives information to the student about their process or strategy for solving tasks and understanding concepts.
- **Specific:** The feedback is targeted and clear so students can discern what they understand, what they do not yet understand, and see how to continue their learning.
- **Timely:** The feedback is immediate if in class and relatively soon if following an assessment so the student can immediately apply and learn from the feedback.

Authors Jeane M. Joyner and George W. Bright (2016) also discuss the significance of actionable feedback focused on student work (and not the student). Such feedback is descriptive, understood by students, and acted on for continued learning growth. It is the type of feedback that students use to monitor their learning.

Much of the feedback and action taken on feedback is the same in Tier 1 and Tier 2 instruction. Both are targeted and descriptive to move students toward learning. However, there are some distinctions for giving descriptive feedback in Tier 1 and Tier 2 (see table 10.2).

There are many aspects to consider when integrating feedback during instruction and after each assessment to build student confidence, understanding, and perseverance. Consider how students give one another feedback through discourse (see chapter 9, page 129, for more information about student discourse), how to tie feedback to essential learning standards and success criteria so students make sense of the feedback (see chapter 11, page 153, for more information about student investment), and how to focus feedback on refining mathematics learning to apply it from one concept to the next.

Also, consider the classroom culture, particularly creating a safe space for students to make and learn from mistakes (see chapter 2, page 27, for more information about classroom culture strategies). Feedback plays a crucial role in students' mathematical learning and their ability to make connections. Teach students how to understand and learn from feedback, as well as how to provide feedback to their peers to accelerate learning to meet grade-level expectations and beyond.

Table 10.2: Use Meaningful Feedback for Learning in Tier 1 and Tier 2

| Use Meaningful Feedback for Learning in Tier 1 and Tier 2 ||
|---|---|
| <ul><li>Feedback is FAST (fair, accurate, specific, and timely).</li><li>Feedback focuses on the student's mathematics thinking.</li><li>Feedback gives information to students about what they have learned or can do, what they still need to learn or know how to do, and insight for next steps.</li></ul> ||
| **Use Meaningful Feedback for Learning in Tier 1** | **Use Meaningful Feedback for Learning in Tier 2** |
| <ul><li>Feedback is generated by the teacher during instruction or by students' peers when working in pairs or groups.</li><li>Feedback focuses on task-level, process-level, and self-regulation observations (see table 10.1).</li></ul> | <ul><li>Feedback is generated by the teacher to clarify any misconceptions or errors in reasoning a student may still have after initial instruction.</li><li>Feedback is task level when students understand the mathematics in the task. If they do not yet understand the mathematics, teaching is required before task-level feedback is helpful. Feedback is also focused on process-level and self-regulation observations (see table 10.1).</li></ul> |

## The *How* of Using Meaningful Feedback for Learning

For students to learn from meaningful feedback, you must create time during lessons and provide feedback in a way that is understood and productive. The mathematics classroom should be like the basketball practice Sarah observed—frequent feedback that is understood and immediately applied. Mistakes are embraced because they are a part of learning, and students begin to evaluate their own progress using feedback.

Descriptive feedback most effectively shows students their current learning and what is not learned yet. Descriptive feedback is not an overall score or generality like saying "good effort," but rather it gives the student information about the quality of their work, which means it is formative in nature. Erkens and colleagues (2017) use the following five questions to guide formative feedback:

1. Does my feedback elicit a productive response?
2. Does my feedback identify what's next for the learner?
3. Is my feedback targeted to each learner's level?
4. Is my feedback strength based?
5. Does my feedback cause thinking? (p. 51)

Additionally, educator and author Nicole Dimich (2024) writes that educators should consider the following to ensure feedback is productive:

- **Be descriptive:** Avoid quantities or general comments like "try again" or "add more." Use descriptive language that tells students about the qualities of their work.
- **Be purposeful:** When providing feedback to improve achievement, your comments should promote learning, not simply justify a score.
- **Begin with a strength and offer a next step:** The strength must be descriptive and relate to the criteria embedded in initial instruction, as should the offered next step.
- **Build in time and learning:** Students need time and instruction to take action on comments.
- **Less is more:** Offer students fewer assignments and assessments but more opportunities to use feedback from teachers and peers to revise and develop their work.
- **Focus on one or two areas, learning goals, or criteria:** Too many comments overwhelm students, and they don't know where to start.
- **Prompt students to action:** Don't fix their mistakes for them. Students are often confused when the teacher writes over their work (fixing it for them), and don't see where the misunderstanding occurred or the process of going from their work to the more quality or advanced work. (pp. 137–138)

To accelerate and grow mathematics learning, use some of the following strategies to strengthen your and your students' feedback practices. These strategies are designed to provide descriptive feedback to students, whether from you, other students, or themselves.

### Show the Mathematics

It is difficult and time consuming to give students feedback during a lesson if their mathematics thinking is not visible in some way. Students might show a graph of a quadratic to find its zeros, show their solution pathway when solving a word problem, or show colored ten frames used to decompose ten. Without the images or work, you must listen very carefully when students try to explain their thinking to make sense of the explanation. Apart from the time it takes to hear students verbally explain their reasoning and make sense of it when there is no work or manipulatives to analyze, the student may not understand the targeted nature of any feedback received. Figure 10.1 shows the difference between hearing student thinking and seeing student thinking.

The feedback for the solutions, whether shared verbally or in writing for each task, remains the same. It might be challenging for a student who only provided a verbal solution to grasp the errors in reasoning or mistakes when given feedback. Can the student discern what is correct in their thinking and what needs to be adjusted? Students may find it difficult to comprehend verbal reasoning and solutions without a visual aid, making it nearly impossible to provide feedback to each other.

Consider the feedback needed for the work shown in figure 10.1. If you want a high school student to understand zeros occur when $y = 0$, not when $x = 0$, you might explain zero is the output or ask the student

| Task | Student Verbal Thinking With an Error or Mistake | Student Written Thinking With an Error or Mistake |
|---|---|---|
| Find the zeros of $y = x^2 + 2x - 3$. | If I plug in 0, I get $0 + 0 - 3$, so the answer is $-3$. | $y = 0 + 2(0) - 3$ <br> $y = -3$ <br> The zero is $-3$. |
| Joe is making cookies. He wants to triple the recipe, which means he needs to triple all the ingredients. The original recipe calls for $2\frac{1}{4}$ cups of flour. He put 2 cups of flour in the batter before he remembered the recipe needed to be tripled. How much more flour does Joe need to add to the batter? | Joe needs to triple $2\frac{1}{4}$ so he needs to multiply $2\frac{1}{4}$ by 3. That means he needs $6\frac{3}{12}$ cups of flour. Since he already put 2 cups into the batter, he needs $4\frac{3}{12}$ (or $4\frac{1}{4}$) more cups of flour. | $2\frac{1}{4} \times 3 = 6\frac{3}{12}$ <br> $6\frac{3}{12} - 2 = 4\frac{3}{12}$ <br> Joe needs to add $4\frac{3}{12}$ (or $4\frac{1}{4}$) cups of flour. |
| Show two ways to decompose ten using a ten frame. | I can make 10 using three red discs and 6 yellow discs or using 5 red discs and 5 yellow discs. | Using a ten frame and colored discs, the student shows and writes: <br> $3 + 6 = 10$ <br> $5 + 5 = 10$ |

**Figure 10.1:** Examples of incorrect student verbal explanations and written work needing feedback.

*Visit **go.SolutionTree.com/MathematicsatWork** for a free reproducible version of this figure.*

to create a graph to see the zeros. You might want a fifth- or sixth-grade student to understand the thinking is correct when solving the task about cups of flour, but there is an error when computing $2\frac{1}{4} \times 3$. Ask the student to use a number line or revisit repeated addition to discover $2\frac{1}{4} \times 3 = 2\frac{1}{4} + 2\frac{1}{4} + 2\frac{1}{4} = 6\frac{3}{4}$.

Perhaps you want a kindergarten student to see they made a simple mistake. Ask the student decomposing ten to show you the 10 for each equation made on a ten frame so they discover that 10 cannot be decomposed into 3 and 6, but rather decomposed into 3 and 7 or 4 and 6 (keeping one number the same as the original answer). When students use their written thinking or visible manipulatives to revise their work with targeted feedback, they can more effectively apply their learning from feedback in the future.

As mentioned in chapter 9 (page 129), mathematics professor, author, and consultant Liljedahl (2021) discusses the importance of students working in random groups of three students in grades 3–12 and two students in the primary grades to solve mathematical tasks. One aspect of Liljedahl's thinking classrooms involves students making their thinking visible on vertical spaces (for example, whiteboards or chart paper on the walls) by working together to show their mathematical reasoning. Students work more productively together when they can see their work and make any adjustments needed. Students are simultaneously generating feedback for and receiving feedback from each other as they work through a given task. Additionally, the teacher can give feedback to students because they can quickly see who is making sense of the task and who might need scaffolded supports since student work is easily visible on the walls.

Some tools that support students in showing their thinking and making mathematics visible in the classroom during lessons include the following.

- Manipulatives (for example, linking cubes, ten frames, fraction tiles, or algebra tiles)
- Whiteboards
- Dry-erase student desks
- Blank paper or chart paper
- Touch screen computers that allow students to record their work using a finger or stylus
- Graph paper
- Calculator or graphing program

Students working in Tier 1 and Tier 2 require feedback. Students are more likely to understand and act on that feedback if their original thoughts and reasoning are easy to see and learn from.

In addition to mathematics tasks posed by a teacher, a book, or a written assignment, computers are also used to generate tasks for students as part of their Tier 1 and

Tier 2 instruction. Unfortunately, too many students think solving problems on the computer equates to using mental mathematics, even when the tasks require multiple steps. They do not see the value in showing the mathematics required for meaningful feedback and learning. We have had success at all grade levels using a piece of folded paper to help students show their thinking when solving tasks from a computer program. Begin by having students fold a piece of paper vertically, then into fourths horizontally. Tell students to write their thinking in each box that shows how they know their answer is correct. Figure 10.2 shows a second-grade example.

The folded paper organizes student work and allows teachers to give students feedback about their thinking.

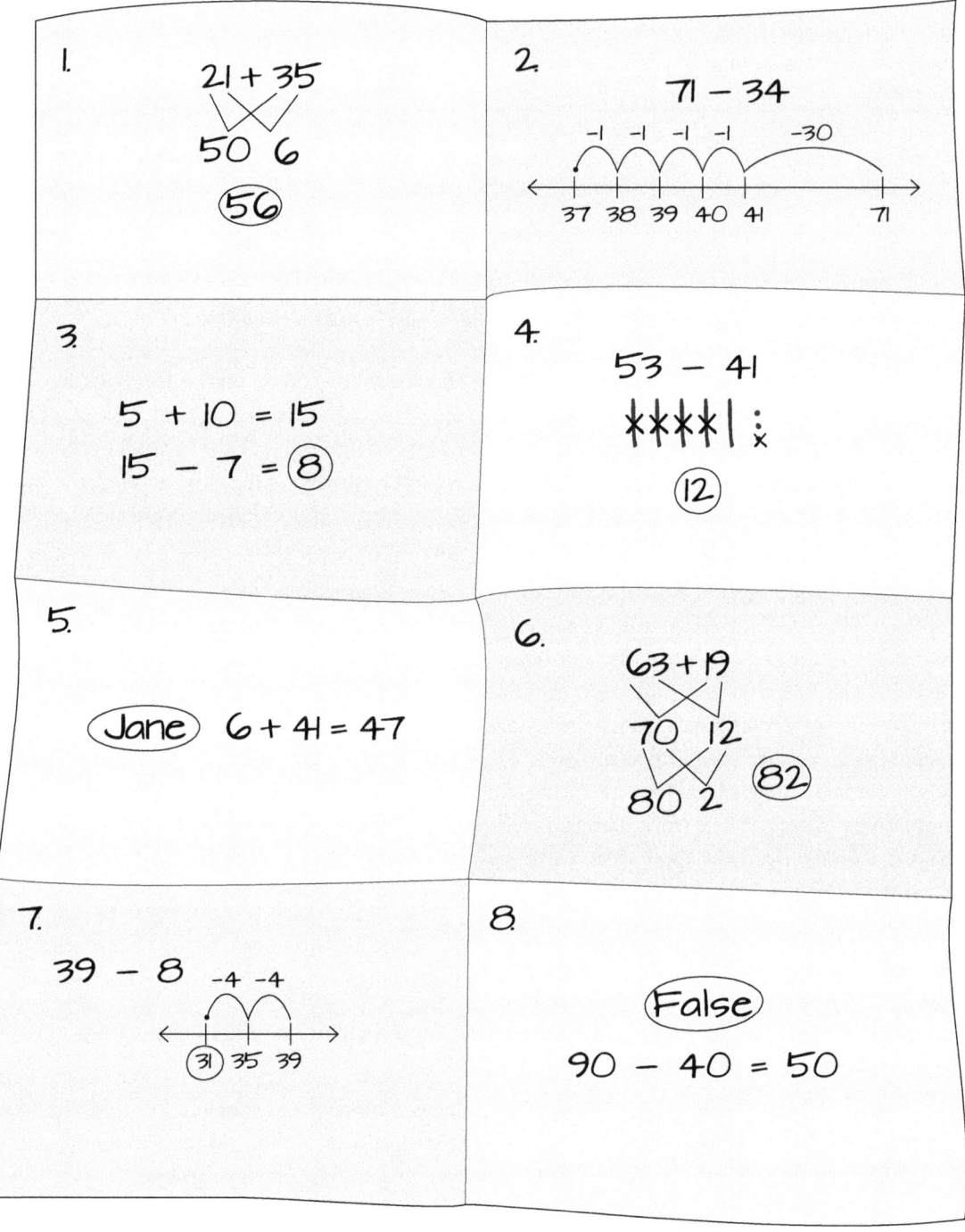

**Figure 10.2: Example of folded scratch paper to make student thinking visible in second grade.**

Students then continue their learning by acting on the feedback and revising their work. Folded scratch paper can be used any time students are completing tasks on a computer, whether during instruction, assessment, or at-home practice.

When students show the mathematics, consider how to increase learning by having them revise their work (recognize their correct mathematics thinking and fix any mistakes) rather than redo their work (start from the beginning and do the problem again without referencing previous work). Consider how to create a classroom expectation that thinking must be visible so students can better give and receive feedback, reflect on their work, see its strengths, and revise their work as needed so their learning can be applied to future tasks.

### Three Es

In *Engage in the Mathematical Practices: Strategies to Build Numeracy and Literacy With K–5 Learners*, Norris and Schuhl (2016) share the Three Es strategy, which involves students generating feedback for themselves by reflecting on the strategies or models used in class to solve mathematics problems. Identify the different strategies or models used after students work individually or in groups to complete a task. Next, select students to post their solutions on a classroom wall so everyone sees them, making sure each strategy used in the class is posted once.

You may want to determine a specific order in which to share student work, from concrete to abstract thinking, or start with a strategy most students used and then share others to make connections. Ask students the following three questions as they look at each solution pathway posted.

- Which of these strategies are *easy* to understand?
- Which of these strategies are *effective*?
- Which of these strategies are more *efficient*?

Students might make sense of adding 35 + 56 by writing 35 dots and 56 dots, or solve a system of linear equations by guessing and checking. Though time consuming and laborious, their choice of strategy may be *easy* to understand and *effective*, if done carefully. However, when elementary students see peers drawing base ten blocks or using place value to add, or high school students see peers graphing equations or using algebra to solve a system of equations, they begin to realize there are more *efficient* ways to solve tasks.

Figures 10.3 and 10.4 (page 146) show examples that students could use with the Three Es strategy to generate their own feedback and determine an efficient way to solve problems.

There is not always one correct or most efficient strategy, but for students not yet using an efficient strategy or model, they can see another option and give it a try. Also, sharing student work and having others reflect on strategies and models helps students see there is more than one way to answer a mathematics problem or task. Students not yet at grade-level understanding make connections to prior learning and see there are different ways to reason to an answer. In the future, they might be more inclined to try something that is efficient and allows for meaningful feedback.

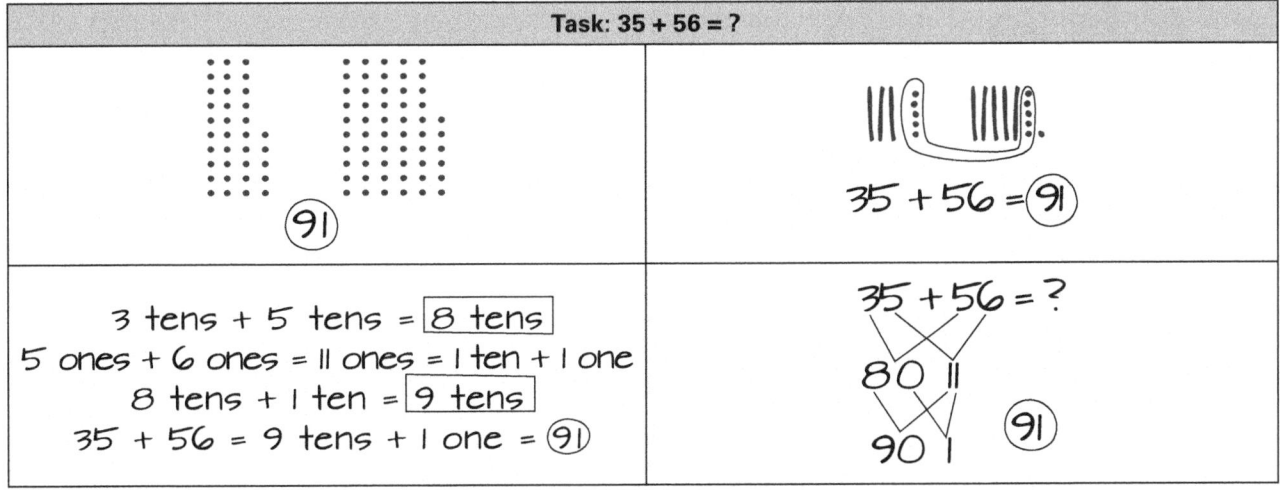

**Figure 10.3: Example of student work showing addition using the Three Es strategy.**

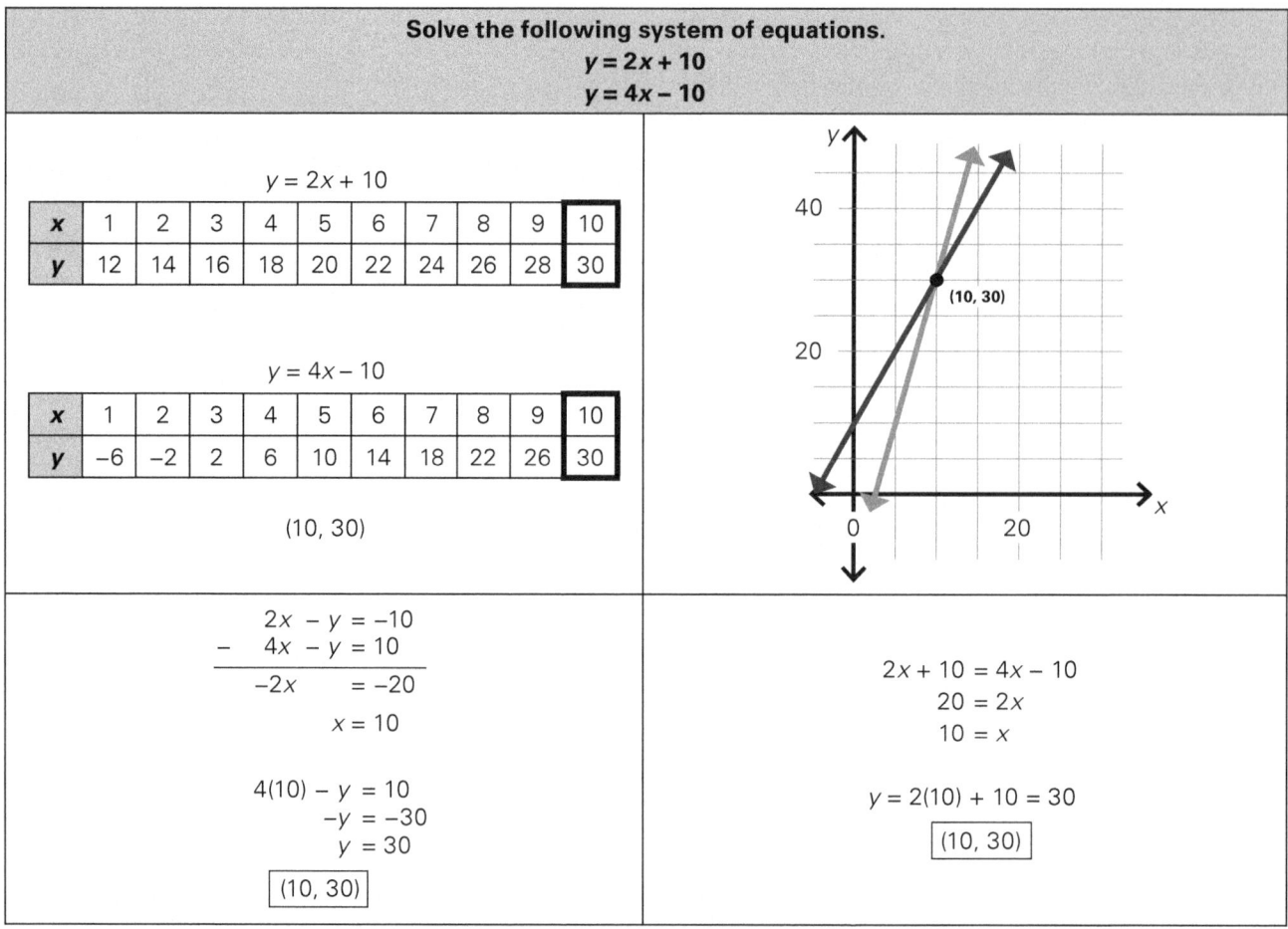

Figure 10.4: Example of student work showing how to solve a system of equations using the Three Es strategy.

You might want to make an anchor chart or poster showing the Three Es strategy to remind students how to reason through solution pathways so students can reference it when doing original work in the classroom (see figure 10.5). Add examples of each *E* to the anchor chart to give students a visual when reflecting on their own work.

The Three Es strategy helps students evaluate their own thinking when solving a task or after it has been completed. Along with giving feedback to classmates, students utilize self-regulation feedback by analyzing their own work in comparison to other strategies generated in the classroom. The Three Es works well in Tier 1. For Tier 2 interventions, if students are not yet solving tasks using multiple strategies, you may want to create and share the different strategies or models yourself and have students use your examples to determine whether their strategy is *easy* to understand, *effective*, and *efficient*.

| Is this strategy EASY to understand? | Is this strategy EFFECTIVE? | Is this strategy EFFICIENT? |
|---|---|---|
| Ask:<br>• Does my strategy make sense?<br>• Can I explain my strategy?<br>• Can I use my strategy again? | Ask:<br>• Does my strategy work?<br>• Is it easy to make a mistake with my strategy?<br>• Do I like using my strategy? | Ask:<br>• Does my strategy take less time to use?<br>• Would I want to use my strategy again? |

Figure 10.5: Three Es anchor chart.

### Provide the Answers

In many mathematics classrooms, the answers are kept secret from students until the task is completed or class is over. Unfortunately, many students share the view that the goal of any mathematics class is to finish quickly and be done. For example, students might say, "I'm done. Can I read a book?" or "I'm done. Can I do my science homework?" Students do not necessarily see the connection between mathematics and their world; instead, mathematics is something to get through and, once the worksheet or task is completed, be finished with for the day.

When you keep answers from students until they have finished—whether the answers are needed for an activity in class or homework practice—students tend to work more quickly and do not check their work against the answers for the purpose of learning. However, if you provide students with the answers, the focus is not on students quicky getting to an answer. Instead, the focus is on showing *why* the answer is correct. Students learn to justify their solution pathway and reasoning that leads to the given answer. Students receive instant feedback on their work because they know immediately whether their work leads to the correct solution. And, if their work is not correct, they are more apt to continue working until the answer matches, having learned from process and self-regulation feedback along the way.

Provide the Answers works best as a feedback strategy with tasks that are not rote in nature (memory tasks like quick recall of addition facts, division facts, and so on; see Just Right Numbers, page 105, for how to use answers to grow procedural fluency). Additionally, the task needs to be solved without simply substituting the answer into the original problem to show it is correct. For example, giving the answer of 6 to $x + 12 = 18$ means students can substitute the 6 into the equation to show it is correct without reasoning to find 6. Students must show how they got from the task to the answer. Figure 10.6 shows some examples of appropriate tasks across grade levels.

Giving the answer at the start of a task does not mean you're taking away student thinking—rather it is quite the opposite. Giving the answer means students are showing their reasoning. If you give students homework, providing the answers allows them to check their reasoning and make as-needed adjustments before independently practicing several problems using incorrect mathematical reasoning.

Giving students the answer also provides the opportunity for them to check their work or thinking midway through a problem rather than only at the end. Students ask themselves or others, "Am I on the right track toward the answer?"—a question mathematicians ask themselves as they work to solve problems.

There are many ways to give feedback by sharing answers with students.

- Write the answers on the board.
- Write answers on folded paper and tape the paper to the wall. Then have students get up and check their answers as they are working or when they finish a task.

| Grade | Task | Answer |
|---|---|---|
| Grade 1 | Maria has 6 pieces of gum. Her friend Joe has more pieces. Altogether, they have 15 pieces of gum. How many pieces of gum does Joe have? | 9 pieces of gum |
| Grade 5 | Tonya ran $3\frac{1}{2}$ miles Monday morning and $4\frac{3}{8}$ miles Monday afternoon. On Tuesday, Tonya ran $8\frac{1}{4}$ miles. Did Tonya run more total miles on Monday or Tuesday? How many more miles did she run? | Tonya ran $\frac{3}{8}$ miles farther on _____. |
| Grade 7 | Javier bought a sweater that was originally $57 and was on sale for 30% off the original price. He had an additional coupon for 25% off any purchase at the store. How much did Javier pay for the sweater? | $29.93 |
| Geometry | Corn is stored in a silo shaped like a cylinder. Farmer John wants to know how much shelled corn the silo can hold. Each cubic foot of the silo can hold 45 pounds of shelled corn. John knows the silo has a height of 65 feet and the circular base has a diameter of 20 feet. About how many pounds of corn can the silo hold? | The silo can hold about 453.5 pounds of corn. |

**Figure 10.6:** Examples of mathematics tasks appropriate for the Provide the Answers strategy.

- On a worksheet of eight problems, for example, write eight answers randomly at the bottom of the sheet and have students check off the correct answers as they find them.
- Use answers in the back of textbooks.
- Show students how to use answer banks as resources to check answers and work their way to the solution.
- Reveal answers on the board (or screen) once students have started to work on the task.

Provide the Answers is a strategy designed to give students immediate feedback and emphasize the need for them to reason mathematically and show their thinking.

### Learning Teams

When you group students together for classroom work, consider how they will interact as a learning team. A learning team is responsible for solving tasks together, supporting one another in learning, and giving feedback to one another. Learning teams can be used for both Tier 1 and Tier 2 instruction.

In chapter 9 (page 129), we discussed student discourse and flexible grouping. You can randomly place students into learning teams using the flexible grouping strategy. Students are expected to engage in discussions within their learning team to solve tasks and provide feedback to each other. Students may need to utilize sentence frames (see chapter 9) as they learn to communicate their mathematical thinking and give feedback. The name *learning team* conveys to students the importance of learning from one another and highlights that everyone on the team is a learner.

### Task Sorting

Another way for students to learn as a group is to sort tasks together. Provide three examples of student work for a task that they have completed (whether created during class or not). Remove students' names from the tasks and paste each sample of work on a colored piece of construction paper before handing them out. For example, paste student sample 1 on pink paper, student sample 2 on white paper, and student sample 3 on purple paper. As a class, discuss what constitutes a strong solution to the mathematics task. Let students sort the task solutions in their groups from strongest to needing the most support.

At the elementary level, you may want to start using just two tasks and the words *glows* and *grows*. When teams have sorted the student work, ask which is strongest. Students hold up the strongest sample, and if they are all holding up the same color, you know they understand the success criteria and can move on to the next sample, always asking, "How can this solution be made even stronger?" If students hold up different colors, you know more discussion is needed for what constitutes strong work.

At the conclusion of the activity, if you used a task that students previously solved in class (such as a class assignment or exit ticket), return each student's work and ask them to improve it using the feedback they generated during the activity. You can suggest using colored pencils or pens to highlight the changes so students can see the difference between their original responses and their stronger, revised responses. Feedback obtained during the Task Sort will help students apply self-regulation feedback and grow an understanding of what they have learned or not learned yet.

### Highlighters

Using highlighters is a great a way to see students' thinking and give targeted feedback. There are several ways that highlighting works as feedback.

- Highlight student work where there is a reasoning error and return it to the student to revise.
- The student highlights a part of their solution they are unsure of when turning in the task or assignment (or before sharing it with a partner or group), which allows you (or other students) to give specific feedback and grow their ability to accurately self-reflect.
- If using a grading rubric, have the student highlight in yellow the level at which they think their work scores. You can then highlight in blue the actual scoring on the rubric. The rubric highlights turn green if they are a match and stay yellow and blue if there is not a match. Students and teachers can use this feedback to promote student learning.

Highlighting is a strategy also used for reading. Help students make the connection between how they use highlighting in literacy and mathematics or other subjects with informational text.

### Student Questions

In chapter 9 (page 129), purposeful questioning was a strategy for developing meaningful student discourse. Purposeful questions are critical to student learning in the classroom (NCTM, 2014). When students learn to generate the questions, they give feedback to themselves or get feedback from the teacher or their peers.

Periodically, you might ask students to share a question they have about the lesson so far related to the learning target for the day. Students share their questions as a think-pair-share, write their question on paper or in a journal, or type their questions into a digital document. Figure 10.7 shows examples of student questions across grade levels.

Ask students to revisit their questions later in the lesson and see if they or their peers can now answer their questions, or if they now have a new question. Student-generated questions give students a chance to pause and reflect on their learning, and the questions give feedback both to you about student learning and to the students about their own learning. (See chapter 9, page 129, for additional questioning strategies to support student-to-student discourse.)

### Error Analysis

Students often think that finding an error in real or fabricated student work is a game and more doable than simply doing the task themselves. We mentioned error analysis in chapter 2 (see pages 36–38) as part of building a learning culture and in chapter 7 (My Favorite Mistake, page 106) to develop procedural fluency in specific tasks. Now, consider how to use error analysis for feedback.

Give students real or fabricated work that has an error, and ask them to work in teams to identify it. As students identify the error and then work to revise the solution, they are simultaneously learning how to give feedback more effectively to one another and themselves.

In a seventh-grade classroom Sarah once observed, the teacher asked students to solve an order of operations task during a class game. The task was: Simplify $100 + 20(3 - 6) \div 2 - 45$. Students simplified the expression and then, using clickers, chose the correct answer. The front board then showed bar graphs for how many students chose each of the four options. The teacher celebrated that most students chose the correct answer of 25 and noticed several students also chose a common incorrect answer, –25. She asked students to figure out the mistake someone would have to make to get the incorrect answer of –25. Students eagerly leaned in, and one student held up his paper and said, "I got –25! Let's look at this to figure it out." He was praised by his team for having the incorrect answer, and they were able to find the mistake!

Errors in student work are opportunities for students to learn. Giving students a worked problem with an error and having them find the error is a way for them to practice giving feedback to one another and themselves. They need to identify the error, communicate it, and then solve the problem correctly.

| Lesson Topic | Examples of Student Questions |
|---|---|
| Kindergarten: Subitize and write the numbers 0–10. | • How do you know how many dots there are without counting them?<br>• How do I make a 3? |
| Grade 3: Represent fractions on a number line. | • If I have to make fifths, do I need 5 tick marks between 0 and 1 or 5 spaces between 0 and 1?<br>• How do I show fractions greater than 1 on a number line? |
| Grade 7: Solve percent problems. | • Can a percent be greater than 100%?<br>• Is 20% off the same as paying 80% of the price?<br>• If I take 5% off, and then use a coupon for an additional 10% off, is that the same as 15% off? |
| High School: Graph a quadratic function. | • Does a parabola always have at least one $x$-intercept?<br>• How can I use the symmetry of the parabola to graph the quadratic function? |

**Figure 10.7:** Examples of student questions across grade levels.

*Visit* **go.SolutionTree.com/MathematicsatWork** *for a free reproducible version of this figure.*

### Learning Goals and Reflection

Feedback provided by the teacher, student peers, or students themselves is more meaningful when connected to learning targets and when students have a clear understanding of the success criteria needed to learn the standard.

Students can write a learning goal and then use the feedback from their classwork and from common assessments to reflect on their progress toward achieving that goal. For example, they might set a goal of earning a 3 (proficient) on the target *I can add fractions with unlike denominators*. Students use their classwork and assessments to see evidence of their learning and track their progress. If they reach their goal, they celebrate, and if they do not meet their goal yet, they analyze their evidence to identify strengths and specific areas they still need to learn. We explore using targets and goals for student feedback in chapter 11 (page 153).

## Conclusion

There are many ways your team and students can generate feedback. Bill Gates once said, "We all need people who will give us feedback. That's how we improve" (BrainyQuote, n.d.). It is not always easy to learn from feedback—in fact, sometimes it means you must acknowledge that an instructional strategy you implemented did not work. For students to learn from feedback, they need to feel safe in their classroom community, which embraces errors as learning opportunities. Also, it is critical for the focus of any feedback to be on the task or process, not the student.

Feedback is part of almost everything that happens in the classroom. Consider how to gather information about student reasoning in your lessons and as a team to inform future instruction in Tier 1 or Tier 2. Additionally, consider how to continuously utilize descriptive feedback as a learning strategy for students engaged in Tier 1 and Tier 2 learning experiences.

## Questions to Consider for Next Steps

As a collaborative team, use the following questions to reflect on your current practices and determine any next steps related to your teams' use of feedback.

1. What type of feedback helps you grow your teaching practices?

2. How do you use feedback related to student learning during lessons?

3. What strategies do you currently use to give students meaningful feedback?

4. How do students learn from feedback you or their peers provide in class?

5. What is a challenge when you have students give feedback to each other or themselves?

6. Which of the strategies in this chapter do you already use? Share how you use this strategy.

7. Which of the strategies in this chapter do you want to try to strengthen the use of feedback in your classroom?

8. What is the next step for you and your team related to meaningful feedback?

Teachers often utilize formative assessment information to understand students' needs, adjust instruction, and deliver targeted learning experiences. In fact, the practice of using results to drive instructional decisions is exactly what the teams invested in at Mason Crest Elementary. Teachers understood their students had diverse learning needs and met routinely to collectively address those needs. However, there was a missing piece to their instructional efforts—the students. The teachers noticed students were only passively involved in understanding their next learning steps. Though the teachers had a clear vision of instruction and delivered differentiated and targeted learning, there was no systematic structure for students to get feedback, visualize next steps, act on feedback, or commit to a goal.

Things changed at Mason Crest Elementary when the kindergarten team used their collaborative team time to create goal-setting cards and develop routines for students to use them. The kindergarten team engaged in active research about the learning progressions for essential standards, carved out time, and created concrete avenues for students to reflect on what they were learning and what they still needed to learn. Kindergartners were able to verbally share what they were learning and actively engage in achieving their next goal, which transformed how their teachers and other teachers within the school viewed formative feedback and action.

If five- and six-year-old students could make sense of and act on feedback, the teachers began to wonder, "Isn't it possible to create avenues for *all* students to do the important work of goal setting?" The teachers set out to determine how to include students in their own learning and goal setting in every grade.

CHAPTER 11

# Empower Learners Through Student Investment

> At the center of the assessment process is the student . . . an important goal of assessment should be to make students self-assessors, teaching them how to recognize the strengths and weaknesses of past performance and use them to improve their future work.
>
> —National Council of Teachers of Mathematics

The teachers and teams at Mason Crest Elementary worked in effective collaborative teams focused on gathering information about student learning (see story on page 152). They planned targeted interventions and differentiated core instruction. Yet, it was not until they created tools for students to set goals and monitor their learning that students fully became part of the learning process. Students understood their progress along a learning continuum for each essential standard and had clarity about what they were learning and what mastery looked like. Students were invested in their learning; they knew what they were learning in Tier 1 and could use that information to understand why they might need additional Tier 2 intervention or extension.

Student investment involves empowering students to take ownership of their learning. This means they can articulate what they have learned or not learned yet and set goals for that learning. They see progress when they focus, persevere, and build a positive mathematics identity. School is no longer done *to* students, but rather *with* students. Just like the teachers at Mason Crest Elementary, your collaborative team can involve students in clarifying and understanding their mathematics learning throughout their academic career.

## The *Why* and the *What* of Empowering Learners Through Student Investment

Assessment experts Erkens and colleagues (2017) describe *student investment* as follows:

Student investment is not about getting students to be compliant; it is about developing students' ability to reflect on their learning in light of a clear learning progression, track their progress, and develop a process for persisting through struggle and growth to achieve more. (p. 113)

Dimich (2014) shares further that students are invested when they do the following:

- Have language to describe their learning
- Have a clear idea of quality and not-so-quality work
- Take action on descriptive feedback
- Revise their work
- Self-reflect on what the assessment means in terms of their learning
- Set goals based on assessment information
- Make an action plan in partnership with teachers to achieve their goals and improve
- Share their work and plans to improve
- Share their thoughts on what helps them learn and what gets in the way of learning
- Experience the ways in which the learning is relevant and challenging through assessments, instructional activities, and homework that teachers design (p. 11)

The work of Erkens and colleagues (2017), as well as Dimich (2024), demonstrates that student investment requires clarity about what is being learned, an understanding of quality work, analysis of one's own work, reflection, and action. Through investment, students see relevance in their learning, which encourages them to continuously work toward grade- or course-level learning and beyond. They develop self-efficacy and persevere.

When students think or say, "I'm bad at math," learning mathematics might feel overwhelming. However, when students can articulate what they have learned and what they have not learned yet, learning mathematics becomes doable because they see the pathway on the learning progression. One way for students to invest in their learning and see their progress is through academic goal setting. Marzano (2023) describes *academic goal setting* as a process that involves teachers supporting students to understand their learning needs and provide them with opportunities to take specific steps toward those goals.

In 1933, educational reformer and philosopher John Dewey recognized the importance of reflection on learning when he stated, "We do not learn from experience. We learn from reflecting on experience." He believed people attached meaning and relevance to experiences by reflecting on them. Jan Chappuis and Rick Stiggins (2020) from the Assessment Training Institute add, "Engaging students regularly in noticing and reflecting on their own progress helps them develop an inner dialogue of self-monitoring necessary to becoming self-regulated learners" (p. 363). Student investment provides learners the time to reflect on their progress, the opportunity to continue learning, and the ability to develop an inner dialogue needed for mathematics agency.

Student investment happens throughout daily lessons and after common assessments. Students learn what proficiency means for each essential standard and use evidence from their own classwork or common assessments to determine their progress. Asking students to reflect on their learning related to the daily target supports their understanding of what they must learn to be successful (Schuhl et al., 2024). Through self-reflection, students will be better equipped to accurately set learning goals, self-assess their progress toward meeting those goals, and create a plan of action when they have errors or misconceptions demonstrated on common assessments (Brown & Ferriter, 2021).

NCTM (2014) recognizes the importance of using assessments to help students identify strengths and weaknesses in their work and improve from that knowledge, thus self-monitoring their learning. There are many opportunities to invest students in their own learning when you are clear about the learning targets and the necessary success criteria.

Students' analysis of their progress and reflection of their learning often does not come naturally. Therefore, you and your team teach, model, and reinforce self-reflection routines so students can determine specifically what they have learned, what they are still learning, and what they have not learned *yet*.

When teaching students to self-regulate and self-reflect, consider the reflection tools your team will use or create. These student reflection tools or processes should be used consistently across your team as a learning routine. When students reflect on their learning and articulate what they have learned or are still learning, they more effectively make connections between Tier 1 and Tier 2 learning experiences. Consider the scenario in figure 11.1 as your team determines how to connect assessment feedback and Tier 1 and Tier 2 instruction.

It is essential for you and your team to provide students with targeted, specific feedback to encourage reflection and build student investment. As a collaborative team, discuss the routines you agree to use to promote this investment. Work collectively to shift student thinking from feedback as a grade to feedback as useful information about learning. Students can track their own learning over time with targeted feedback.

Employ student investment strategies during Tier 1 using quality feedback throughout lessons and after common assessments so students know what they learned and what they have not learned yet. Students use this knowledge to inform the learning needed in Tier 2 interventions or extensions. Table 11.1 shows additional distinctions between your instructional student investment practices in Tier 1 and Tier 2.

During Tier 1, your collaborative team designs lessons tied to clear learning targets, actively engages students in understanding what they are learning, helps students monitor their progress, and sets goals for the class. During Tier 2, re-engage students in learning and help them see their progress through continued reflection on the same learning targets used in Tier 1. Next are strategies for how to involve students in their learning through student investment.

**Directions:** Read the scenario and discuss the questions as a collaborative team.

**Scenario:** Chase received his unit 1 assessment results. Overall, he earned a score of 63%. Two teachers on the team provided different feedback. The following feedback shows that teacher A gave Chase a single score and Teacher B gave Chase feedback by target.

**Feedback From Teacher A**

| Sixth-Grade Student: Chase | Number of Correct Items |
|---|---|
| Unit 1 Test | 19/30 |
| Overall Score | 63% |

**Feedback From Teacher B**

| Sixth-Grade Student: Chase | Number of Correct Items |
|---|---|
| I can organize data collected from a survey question into a chart or graph. | 10/10 |
| I can calculate measures of central tendency from a data set. | 8/10 |
| I can interpret my data. This means I can draw conclusions about data. | 1/10 |
| Overall Score | 63% |

**Discuss:** When Chase goes to Tier 2 intervention time, which feedback will clarify the language to use when describing what he has learned or not learned yet? Which feedback will encourage Chase to more effectively self-reflect on his learning?

**Figure 11.1:** Feedback for student reflection scenario and discussion.

*Visit **go.SolutionTree.com/MathematicsatWork** for a free reproducible version of this figure.*

Table 11.1: Empower Learners Through Student Investment in Tier 1 and Tier 2

| Empower Learners Through Student Investment in Tier 1 and Tier 2 |
|---|
| • Use formative assessment data during lessons and from common assessments to determine student learning needs and communicate progress to students.<br>• Connect student self-assessment actions in Tier 1 to Tier 2 re-engagement. |

| Empower Learners Through Student Investment in Tier 1 | Empower Learners Through Student Investment in Tier 2 |
|---|---|
| • Agree on the feedback process as a team, including the structures and pathways for students to reflect on their learning, set goals, and actively pursue those goals.<br>• Teach, model, and reinforce routines for student self-reflection and goal setting. | • Have students monitor their progress during Tier 2 and connect new learning to their previously set goals.<br>• Have students track their new learning.<br>• Track student learning as a team and plan varied re-engagement strategies to support continued learning. |

# The *How* of Empowering Learners Through Student Investment

Strategies to support student investment in mathematics can be applied in all grade levels, during both Tier 1 and Tier 2 learning experiences. Exactly how these strategies are implemented and managed is flexible based on the essential learning students are working to achieve. This chapter highlights a few strategies used to invest students in their learning so they see progress toward learning grade- or course-level standards.

## Goal Cards

It is impossible for students to hit a learning target they cannot see (Chappuis & Stiggins, 2020; Marzano, 2017). Therefore, it is important to help students see the targets so they know what they have learned and what they still need to learn. In the chapter-opening scenario on page 152, the Mason Crest Elementary kindergarten team created goal-setting cards and developed routines for students to use the cards. The teachers developed clear learning progressions on the essential standards to share with students during Tier 1 instruction.

Creating goal cards showing a learning progression not only gives students a clear pathway to success, but also clarifies learning for you and your team. Tools that communicate the learning needed to reach mathematics goals may also help families understand their children's progress and needs.

Figure 11.2 shares an example of a student goal card showing a learning progression for addition and subtraction of whole numbers.

Primary students need specific guidance to effectively use a goal card like the one in figure 11.2. Since many of the strategies on the addition goal card must be observed, primary learners can get feedback during, or immediately following, each assessment during small-group instruction. The teacher asks questions related to the goals such as, "What strategy did you use? What part of the goal card can you do now? What is your next goal? Are you excited for the next step?" The goal card's visuals aid students in recognizing exactly where they are in their learning. With teacher guidance, primary students are explicitly taught what it means to reflect on learning and identify what to work toward next.

The goal card example for intermediate and middle school students in figure 11.3 takes a slightly different approach. Rather than focusing on a progression of strategies, it shows the content pathway toward multistep problem solving with fractions using all four operations. It demonstrates how learning targets build on each other, and it helps students see what is required to fully learn a standard. Since assessment occurs in ongoing ways through task completion, exit tickets, mid-unit quick checks, and more, students can work toward each target at varied paces and receive feedback along the way.

Similar to figures 11.2 and 11.3, figure 11.4 shows an example of a high school goal card with a learning progression students might use for self-reflection and goal setting when learning to solve right triangles.

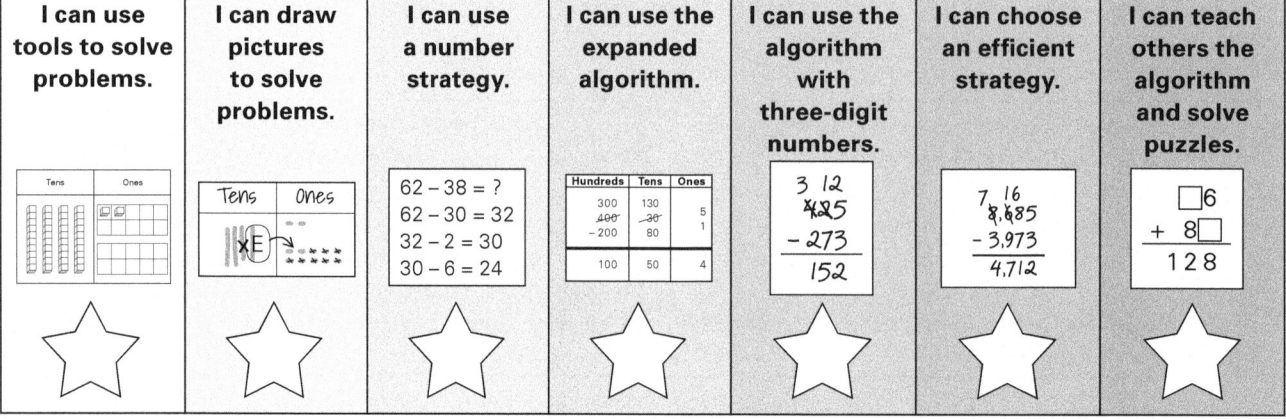

**Figure 11.2: Student goal card example—adding and subtracting whole numbers.**

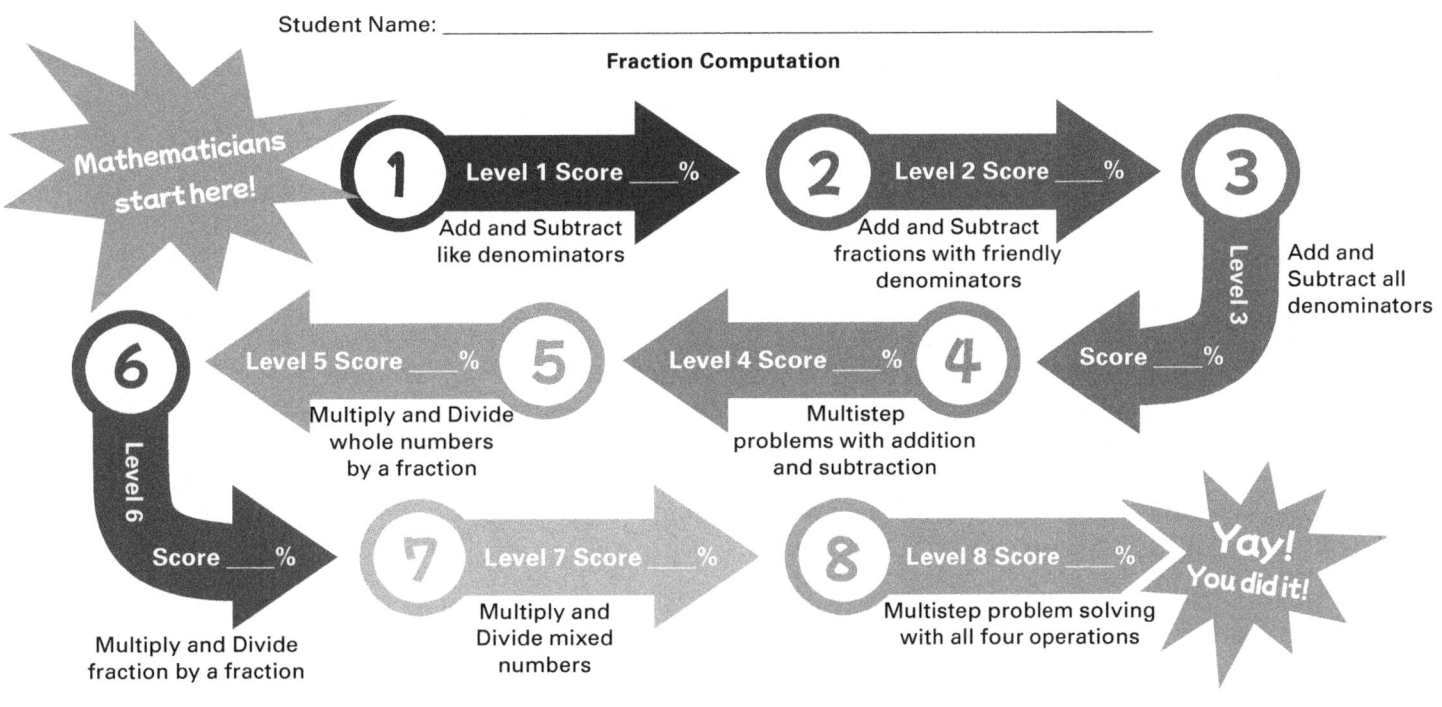

Figure 11.3: Student goal card example—fraction computation.

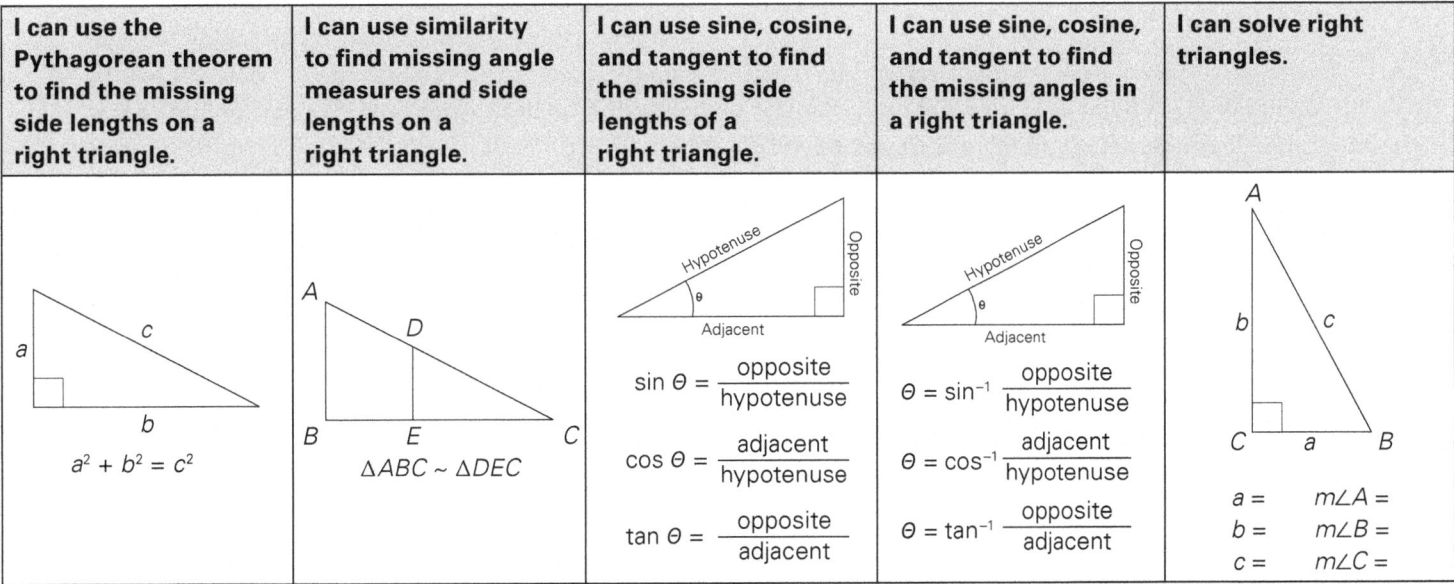

Figure 11.4: Student goal card example—solving right triangles.

You and your team can use goal cards—like those shown in figures 11.2–11.4—to engage students in whole-group discussions about establishing expectations for achieving learning targets. During Tier 1, intentionally connect the targets on the goal card to the strategies and problem-solving techniques learned during both whole-group and small-group settings. Refer to the targets on the goal cards as students work and discuss their learning. These conversations will foster celebrations, genuine peer support, and a collective emphasis on learning.

You and your team can create goal cards using learning progressions to reach each grade- or course-level essential standard. You can use online resources, such as the coherence maps at Achieve the Core (https://tools.achievethecore.org/coherence-map), to determine a learning progression. By building shared knowledge of strategy or content progressions, you can gain clarity on the step-by-step targets of a goal-setting template and promote more targeted learning in both Tier 1 and Tier 2.

### Goal Setting

Students receive feedback after each common assessment, which shows their current level of learning. In *Mathematics Assessment and Intervention in a PLC at Work, Second Edition,* Schuhl and colleagues (2024) show how to design meaningful common assessments and give feedback to students by learning target, like Teacher B in figure 11.1 (page 155).

When assessment scores are given by target, students can review their feedback, learn from their mistakes or errors in reasoning, and set goals for continued learning. Students often set goals for future assessments based on evidence of their learning on recent common assessments. You and your team can ask students to reflect on the questions in figure 11.5 once they receive their assessment scores by target.

Figure 11.6 shows an assessment reflection from Mission Oak High School. Students chose a plan for continued learning before taking a reassessment and then reflected on their learning from that reassessment. Their teacher provided feedback on their reflection and plan.

Student investment is an ongoing process during an instruction unit. Students regularly set goals, reflect on their progress, and analyze their work during each lesson and following each common assessment.

### Rubrics

Rubrics are another effective strategy for communicating learning goals to students. Rubrics are evaluation tools and guidelines that share leveled learning expectations. They clarify what students are learning during Tier 1 instruction and how to demonstrate that learning (Lang-Raad & Marzano, 2019). Rubrics make the grade- or course-level learning transparent to students, families, and your team.

Research suggests using rubrics positively impacts students' self-regulation and self-efficacy (Smit et al., 2017). Your team uses rubrics to give students feedback, and students use rubrics to evaluate their own work and create an improvement plan, as well as give feedback to their peers. There are three different types of rubrics your team might use to grow student investment: (1) holistic rubrics, (2) analytic rubrics, or (3) single-point rubrics.

| **Directions:** Use your common assessment results to identify the learning targets you have learned and those you have not learned yet. Determine a plan to learn those targets. ||
|---|---|
| **Learning Targets I Know and Can Do** | **Learning Targets I Have Not Learned Yet** |
|  |  |
| **My plan to learn is . . .** ||
|  ||

**Figure 11.5: Common assessment reflection example.**

*Visit go.SolutionTree.com/MathematicsatWork for a free reproducible version of this figure.*

| LT5c Test Analysis  Name: Student 1 | LT5c Test Analysis  Name: Student 2 |
|---|---|
| Date: 2/21/24  Period: 5 | Date: 2/21/24  Period: 5 |
| What was your quiz score? 45 % | What was your quiz score? 24 % |
| What did you do after the quiz to prepare for this test (select all that apply)?<br>☐ Completed my quiz evaluation in detail<br>☐ Asked questions in my group if I didn't understand my quiz mistake<br>☐ Asked my teacher questions if I didn't understand my quiz mistake<br>☑ Completed the practice test to 100%<br>☐ Attended tutoring at lunch or after school<br>☐ Studied for the test using other resources. What other sources did you use? _____ | What did you do after the quiz to prepare for this test (select all that apply)?<br>☑ Completed my quiz evaluation in detail<br>☐ Asked questions in my group if I didn't understand my quiz mistake<br>☐ Asked my teacher questions if I didn't understand my quiz mistake<br>☐ Completed the practice test to 100%<br>☐ Attended tutoring at lunch or after school<br>☐ Studied for the test using other resources. What other sources did you use? _____ |
| Based on how I have prepared since the quiz, I predict that my score on this test will be 82 % | Based on how I have prepared since the quiz, I predict that my score on this test will be 10 % |
| **This section to be completed after you analyze your test scores.** | **This section to be completed after you analyze your test scores.** |
| What is your test score? 100 %<br>What type of common mistake did you make on this test (select all that apply)?<br>☐ Input mistake ☐ Calculation mistake ☐ Conceptual mistake<br>What concepts do you still not understand?<br>*Initial value on the temperature problem*<br>Reflect on how you performed on the test based on your quiz score and your predicted test score.<br>*I performed way better than I did on my quiz. I was ready for the test.*<br>*You're doing excellent work!*<br>What could you have done differently before the test so you didn't make the same mistakes?<br>*Do all the homework for unit 5c.* | What is your test score? 45 %<br>What type of common mistake did you make on this test (select all that apply)?<br>☐ Input mistake ☑ Calculation mistake ☑ Conceptual mistake<br>What concepts do you still not understand?<br>*The concept I don't understand is the percentage on the equation, also known as the rate.*<br>Reflect on how you performed on the test based on your quiz score and your predicted test score.<br>*I didn't really do well on my quiz because I didn't prepare for either of them.*<br>What could you have done differently before the test so you didn't make the same mistakes?<br>*Things I could do different are to study, do all my practices before the test, and ask my teacher to help when I don't understand something.*  *Good plan!* |

*Source: © 2024 by Mission Oak Algebra 2 Team. Used with permission.*

**Figure 11.6 Student test analysis plan examples.**

### Holistic Rubric

A holistic rubric describes levels of proficiency for a grade-level standard by combining all the criteria used to evaluate student learning together. Typical holistic rubrics provide a single score (usually on a scale of 1 to 4) (Center for Teaching and Learning, n.d.). Figure 11.7 shows a holistic rubric for a third-grade mathematics standard.

Use a holistic rubric to check for student understanding throughout or at the end of a lesson when teaching student investment. Students use their class work to identify their current level of understanding for each standard.

Figure 11.8 (page 160) illustrates an adaptation of Marzano's (2006) self-assessment rubric. This is an example of a holistic rubric that students can reference

**Standard:** Draw a scaled bar graph to represent a data set with several categories. Solve one- and two-step problems using information presented in scaled bar graphs.

| 1<br>Beginning | 2<br>Approaching | 3<br>Proficient | 4<br>Advanced |
|---|---|---|---|
| Student reads a scaled bar graph and identifies quantities shown in the graph. | Student draws a scaled bar graph to represent a data set with several categories (may contain a minor mistake) and solves one-step problems using information in a scaled bar graph. | Student draws a scaled bar graph to represent a data set with several categories and uses a scaled bar graphs to answer one- and two-step problems. | Student draws a scaled bar graph to represent data with several categories and compares it to a given scaled bar graph. Student answers questions that involve comparing the two scaled bar graphs. |

**Figure 11.7: Grade 3 holistic rubric example for a mathematics standard.**

| Rating | Description |
|---|---|
| 4 | I can teach this to someone else. |
| 3 | I can do this without any help. |
| 2 | I can do this with a little help. |
| 1 | I do not get it yet even with help. |

*Source: Adapted from Marzano, 2006.*

**Figure 11.8: Self-assessment rubric.**

in the middle or at the end of a lesson to self-assess current understanding. This rubric promotes student reflection when used during Tier 1 instruction.

Holistic rubrics are a starting point for student reflection. However, holistic rubrics, as shown in figure 11.8, do not provide targeted feedback to students. Consider using an analytic rubric for more specific feedback.

### Analytic Rubric

Analytic rubrics evaluate multiple criteria with leveled descriptions for each criterion. An analytic rubric clarifies the quality of student work in a mathematical task, assignment, or project separately for each criterion (Gonzalez, 2014). An analytic rubric allows learners to identify their strengths and areas for each criterion to improve. Figure 11.9 shows an example of an analytic rubric for problem solving in the middle school grades.

Students use an analytic rubric to evaluate their current level of understanding of each criterion during Tier 1 and identify areas that still need to be strengthened. An analytic rubric can be used during Tier 2 when students re-engage in learning one specific criterion, both to clarify the learning needed in Tier 2 and to celebrate growth. The main advantage of an analytic rubric is the specificity in feedback on each criterion. Students use feedback from an analytic rubric to plan next steps.

### Single-Point Rubric

The third type of rubric is a single-point rubric. A single-point rubric is designed to communicate learning targets and give students or teachers the opportunity to provide feedback aligned with each learning target or skill. Students' peers and teachers share strengths in student work using the rubric and guide the student toward improvement without over-emphasizing a grade or score.

| Problem-Solving Components | Proficient 3 | Approaching 2 | Developing 1 |
|---|---|---|---|
| Summarize the question and identify what is known and unknown. | Student can summarize the question, identify the given information, and state what they need to know. | Student can summarize the question and identify the given information or state what they need to know. | Student can summarize the question but cannot identify given information or what is needed to solve the problem. |
| Show work. | Student shows calculations as well as charts, graphs, diagrams, or drawings that illustrate the mathematics. A reader can clearly follow their thinking. | Student shows calculations and may show a chart, table, graph, or appropriate diagram. A reader can follow their thinking, even though there may be some small gaps without reasoning shown. | Student shows a chart, table, or graph, but there are mistakes or the work is incomplete. A reader has difficulty following their thinking. |
| State the answer. | The solution or outcome given is:<br>• Correct<br>• Mathematically justified<br>• Supported by the work | The solution or outcome given is:<br>• Incorrect due to minor errors in the work<br>• Correct but work minimally supports the answer<br>• Partially complete or partially correct | The solution or outcome given is:<br>• Incorrect or incomplete<br>• Correct but conflicts with or is not supported by the work |
| Verify the answer. | The verification justifies the solution or outcome completely by doing one of the following.<br>• Using a different strategy<br>• Checking calculations<br>• Providing context | The verification justifies the solution or outcome partially by doing one of the following.<br>• Using a different strategy<br>• Checking calculations<br>• Providing context | The verification does not justify the work or is missing. |

**Figure 11.9: Analytic rubric for problem solving.**

*Visit* **go.SolutionTree.com/MathematicsatWork** *for a free reproducible version of this figure.*

Unlike analytic or holistic rubrics, a single-point rubric clarifies only what a student must know and do to be proficient with a standard; it does not specify the work beyond or below proficiency. According to educator Jennifer Gonzalez (2014), a single-point rubric is often easier for the teacher to create while simultaneously giving the student higher-quality feedback. Students focus on the grade- or course-level criteria rather than all levels of proficiency for each criterion.

Figure 11.10 shows a sixth-grade single-point rubric for an essential standard. As students are working on a mathematical task or problem set that requires them to complete all aspects of the standard, they use a single-point rubric to evaluate each other's work or their own.

Keep the single-point rubric simple in the primary grades. You and your team might also consider including pictures or mathematics examples to help students understand the rubric. Figure 11.11 (page 162) shows a single-point rubric for first grade.

Remember to use single-point rubrics for targeted feedback. Single-point rubrics also enable students to effectively provide feedback to each other and clarify the work and reasoning needed to demonstrate proficiency.

Once students complete their solution pathways on a mathematical task or assessment, ask them to swap their work with a peer. Each student uses the single-point rubric to provide written feedback on one or both sides of each criterion in the middle column of the rubric. If a student's work is proficient, peers can provide specific evidence to support the middle column. Use sentence frames (see chapter 9, page 129) to help students learn how to provide specific feedback to peers.

### Success Criteria

Another strategy to build student investment is to create success criteria for essential standards. Success criteria show the learning targets for the entire standard. Your team identifies these criteria when you unwrap and make sense of an essential standard. Success criteria specify what students are expected to do to demonstrate learning. They provide a map to the learning destination—the specific actions or skills students need to be proficient.

Figures 11.12 (page 162) and 11.13 (page 163) share examples of success criteria. Consider the essential standard and the differences and similarities between the learning targets and success criteria.

---

Name: _____ Period: _____

**Standard:** 6.RP.A.3 Use ratio and rate reasoning to solve real-world and mathematical problems.

| Not Yet<br>*What I Am Still Learning* | Proficient<br>*Criteria for the Standard* | Advanced<br>*Evidence of Exceeding Standard* |
|---|---|---|
| | I can create tables of equivalent ratios relating quantities with whole-number measurements and find missing values in tables. | |
| | I can use ratio tables to compare equivalent ratios. | |
| | I can plot pairs of values on the coordinate plane and explain their meaning using ratio understanding. | |
| | I can solve rate problems, including unit pricing and constant speed. | |

*Source for standard: NGA & CCSSO, 2010.*

**Figure 11.10: Grade 6 single-point rubric example.**

*Visit **go.SolutionTree.com/MathematicsatWork** for a free blank reproducible version of this figure.*

| Name: _____ | | Period: _____ |
|---|---|---|

**Standard:** 1.NBT.B.3 Compare two two-digit numbers based on meanings of the tens and ones digits, recording the results of comparisons with the symbols >, =, and <.

| Not Yet<br>*What I Am Still Learning* | Proficient<br>*Criteria for the Standard* | Advanced<br>*Evidence of Exceeding Standard* |
|---|---|---|
| | I know the number of tens and ones in a number.<br><br>23 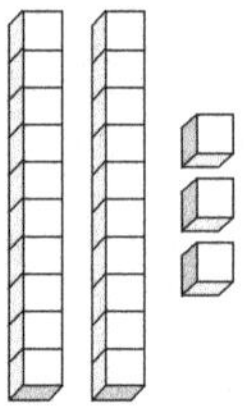<br><br>2 tens + 3 ones | |
| | I can compare numbers using tens and ones.<br>Which is greater? 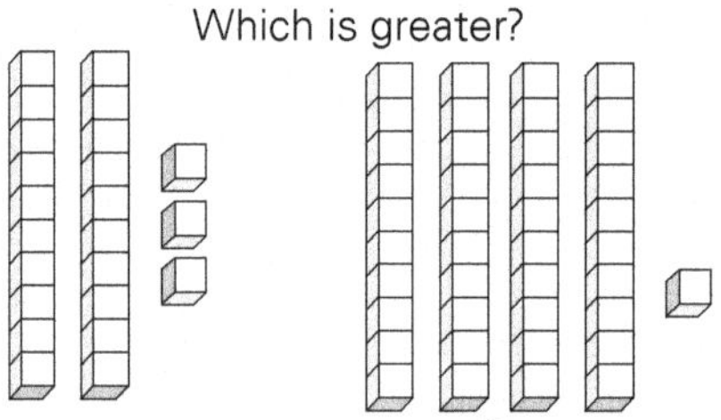<br>23   or   ㊶ | |
| | I can use >, =, or < to write my answer when I compare numbers.<br>23 < 41 | |

*Source for standard: NGA & CCSSO, 2010.*

**Figure 11.11: Grade 1 single-point rubric example.**

---

**Directions:** Read the standard, learning targets, and success criteria. What are the similarities and differences between learning targets and success criteria?

| Essential Standard ||
|---|---|
| **Essential Standard:**<br>2.NBT.A.3 Read and write numbers to 1,000 using base ten numerals, number names, and expanded form. ||
| **Learning Targets** | **Success Criteria** |
| 1. I can read numbers to 1,000.<br>2. I can write numbers to 1,000. | ☐ I can read numbers to 1,000 using place values (for example, 324 is read three-hundred twenty-four).<br>☐ I can write numbers to 1,000 given the number name or expanded form.<br>☐ I can write the number name for a number or expanded form.<br>☐ I can write the expanded form for a number or a number name. |

*Source for standard: NGA & CCSSO, 2010.*

**Figure 11.12: Learning targets and success criteria for essential standard—grade 2 example.**

*Visit go.SolutionTree.com/MathematicsatWork for a free blank reproducible version of this figure.*

| Directions: Read the standard, learning targets, and success criteria. What are the similarities and differences between learning targets and success criteria? |
|---|
| **Essential Standard** |

**Essential Standard:**
8.EE.C.8: Analyze and solve pairs of simultaneous linear equations.
  a. Understand that solutions to a system of two linear equations in two variables correspond to points of intersection of their graphs because points of intersection satisfy both equations simultaneously.
  b. Solve systems of two linear equations in two variables algebraically and estimate solutions by graphing the equations. Solve simple cases by inspection. For example, $3x + 2y = 5$ and $3x + 2y = 6$ have no solution because $3x + 2y$ cannot simultaneously be 5 and 6.
  c. Solve real-world and mathematical problems leading to two linear equations in two variables. For example, given coordinates for two pairs of points, determine whether the line through the first pair of points intersects the line through the second pair.

| Learning Targets | Success Criteria |
|---|---|
| 1. I can solve a system of two linear equations by:<br> • Graphing<br> • Substitution<br> • Elimination<br> • Inspection<br>2. I can explain the meaning of a solution to a system of two linear equations.<br>3. I can estimate a solution to a linear system of equations by graphing.<br>4. I can solve real-world problems using a system of two linear equations. | ☐ I can solve or estimate a solution for a system of two linear equations by graphing.<br>☐ I can solve a system of two linear equations using substitution.<br>☐ I can solve a system of two linear equations using elimination.<br>☐ I can explain the meaning of a solution to a system of two linear equations in context.<br>☐ I can write a system of two linear equations to represent a real-world situation.<br>☐ I can solve a real-world problem using a system of two linear equations. |

*Source for standard: NGA & CCSSO, 2010.*

**Figure 11.13: Learning targets and success criteria for essential standard—grade 8 example.**

Success criteria include the specific terms from the essential standard, as well as clear and concise wording; there is no subjective language. Success criteria also include additional details that might not be explicitly written in the standard but are at grade level and needed to be successful. For example, in third grade, a standard asking students to show fractions $\frac{1}{b}$ on a number line will include success criteria about partitioning the number line into $b$ equal parts between 0 and 1, labeling the tick marks created, and showing the fraction on the number line. As your team is identifying learning targets and writing success criteria, be sure to compare the success criteria with the standard to ensure they accurately match the essential standard's concepts, skills, and intended rigor (Ainsworth & Viegut, 2006; Ainsworth & Viegut, 2015).

### Proficiency Scales

The three types of rubrics and success criteria shared in this chapter are all created to clarify for students what grade- or course-level learning is needed to meet the standard. An additional tool to build student investment is a proficiency scale. Proficiency scales show a learning progression toward grade- or course-level understanding (level 3.0) and beyond. Teachers and students use proficiency scales to evaluate progress toward demonstrating proficiency on an entire standard and inform the types of assessment items needed to measure student learning (Marzano, 2017). Proficiency scales are created and used as tools to provide feedback during instruction and make the essential standard learning progression visible for students.

Proficiency scales articulate the learning progression of a standard in four levels (4.0, 3.0, 2.0, and 1.0). Level 3.0 represents learning a standard to proficiency. Level 4.0 represents a deeper understanding of the standard beyond grade-level expectations. Levels 2.0 and 1.0 show a minimal or partial understanding of the standard. It is possible to include a prior knowledge skill or standard as part of level 1.0.

Marzano (2017) includes half points between levels. For example, 2.5 means a student is close to learning a standard at the proficiency level, but not quite at full proficiency yet. Educational policy professor Thomas R. Guskey (2015) shares that half points on a proficiency scale can be problematic because it is more difficult for teachers to calibrate a student's level of understanding to seven levels (1.0, 1.5, 2.0, 2.5, 3.0, 3.5, and 4.0) than four levels (1.0, 2.0, 3.0, and 4.0). We will use four levels.

When your collaborative team creates a proficiency scale for an essential standard, determine instructionally how to teach the standard. List the targets from least complex (level 1.0) to most complex (level 3.0). For level 4.0, you and your team are doing the early work of addressing critical question 4 of a PLC at Work—defining a collective response to students who are already proficient, which can be used in Tier 2 extensions (Dufour et al., 2024).

Figure 11.14 is a sample proficiency scale using the essential standard in figure 11.13 (page 163). As a team, look at the similarities and differences between the two figures.

There are several advantages to using proficiency scales for student investment.

- Offers consistency of feedback to students across grade and course level
- Clarifies the learning progression of the essential standards
- Creates a visible learning path for students
- Promotes student reflection on their current learning and next steps

Proficiency scales show the learning progression needed to fully master a standard. Rubrics, however, show the level to which a student demonstrates proficiency when answering grade-level questions about a standard (learning level at the 3.0 mark on the proficiency scale). Students see the path to learning with the proficiency scale, and they see the necessary quality of work at grade level to be proficient using a rubric or through success criteria.

| **Essential Standard:** 8.EE.C.8: Analyze and solve pairs of simultaneous linear equations. | |
|---|---|
| | a. Understand that solutions to a system of two linear equations in two variables correspond to points of intersection of their graphs, because points of intersection satisfy both equations simultaneously. |
| | b. Solve systems of two linear equations in two variables algebraically and estimate solutions by graphing the equations. Solve simple cases by inspection. For example, $3x + 2y = 5$ and $3x + 2y = 6$ have no solution because $3x + 2y$ cannot simultaneously be 5 and 6. |
| | c. Solve real-world and mathematical problems leading to two linear equations in two variables. For example, given coordinates for two pairs of points, determine whether the line through the first pair of points intersects the line through the second pair. |
| 4 | • Create a linear equation that intersects a given line at a given point.<br>OR<br>• Determine if a system of three linear equations has a solution.<br>OR<br>• Create a word problem for a given pair of linear equations and then solve the system of linear equations. |
| 3 | • Solve real-world problems by writing and solving a system of linear equations.<br>• Solve systems of two linear equations in two variables using an efficient method.<br>• Identify systems of linear equations with one solution, no solution, or infinitely many solutions. |
| 2 | • Solve a simple system of linear equations using elimination.<br>• Solve a simple system of linear equations using substitution.<br>• Solve a simple system of linear equations by graphing.<br>• Graph a system of linear equations to estimate its solution. |
| 1 | • Estimate a solution to a system of equations from a graph. |

*Source for standard: NGA & CCSSO, 2010.*

**Figure 11.14: Proficiency scale for grade 8 essential standard.**

*Visit **go.SolutionTree.com/MathematicsatWork** for a free blank reproducible version of this figure.*

## Student Trackers

In addition to using goal cards, rubrics, success criteria, and proficiency scales, you and your team can also create student trackers as a reflection tool for students to use throughout an instruction unit and after common assessments. Student trackers include student-friendly learning targets, independent practice, and academic vocabulary for the entire unit and are structured for students to routinely document their learning toward the essential standards in each unit (Schuhl et al., 2024).

Figure 11.15 (page 166) shows an example of a grade 7 student tracker. You can give students a copy of a tracker like this on cardstock at the beginning of the unit. All the work during the unit is aligned to one of the learning targets so students can use the tracker for continual reflection as they learn.

The additional learning opportunities your team creates for students can be listed on the second side of the tracker, which allows students to choose one, if needed, for additional learning (see figure 11.15). Schuhl and colleagues (2024) share the following about using student trackers:

> Because students are clear about what they should be learning, they are able to articulate the standards they need help with as students become part of the intervention process in a productive way. Students work to learn the content with other students in the same course who may have different teachers, expanding their resources for feedback. Additionally, the students become more aware of which mathematics standards they most need to study and learn prior to the common end-of-unit assessment. (p. 111)

Figure 11.16 (page 168) shows an example of a high school student tracker.

To grow student investment throughout each unit, build routines around student self-assessment. Consider how you will require students to use trackers to monitor progress during Tier 1, and then plan for how students will re-engage in learning during Tier 2.

Elementary student trackers include the essential learning standards written as student-friendly targets and may also show common homework (intermediate grades) or a more detailed rubric or scale with sample problems so students better understand the learning target. The trackers may be used weekly instead of for the entire unit.

Figure 11.17 (page 169) shows a second-grade weekly tracker with the learning target for each day and the specific success criteria addressed each day. On the bottom row, students circle the plus sign if they learned and the question mark if they still have a question. You might choose to send student trackers in take-home folders as a communication tool to both the students and their families (Schuhl et al., 2024).

Through your intentional response when students are not learning, and your expectations for them to track their own progress, you signal to students that learning the essential standards for mathematics is of utmost importance and their top priority in your class.

## Student Surveys

Sometimes, your teacher team members are so mired in planning, teaching, and assessing that you might forget to ask for student voice as part of student investment. Consider asking students what you do in class that helps them learn and what gets in the way of their learning. Try to give these surveys at the end of each trimester or semester and analyze the responses as a collaborative team. Consider asking students about their interests so you can bring those into the classroom when problem solving.

Additionally, you might ask students what their favorite learning target in a unit is and why or whether they learn best in whole-group or small-group settings. Student investment includes students being heard and valued in the classroom. As a team, you may specifically want to ask questions related to the Tier 1 and Tier 2 learning experiences offered and how effective and valuable students view them.

Name: _____ Period: _____ Date: _____

**Student Tracker Unit 3: Expressions, Equations, and Inequalities**

**Learning targets:**

1. I can simplify algebraic expressions.  | Starting . . . | Getting there . . . | Got it! |

2. I can write and solve an equation from a word problem. | Starting . . . | Getting there . . . | Got it! |

3. I can write and solve an inequality from a word problem. | Starting . . . | Getting there . . . | Got it! |

4. I can graph the solution to an inequality and explain how the solution answers the question. | Starting . . . | Getting there . . . | Got it! |

**Note:** Teacher will use one or two warm-up questions that relate to each of the preceding learning targets to show student proficiency. Students must complete the following assignments to prepare for the warm-up questions.

**Common Homework Assignments**

| Target Number | Lesson | Page Number | Problem Numbers | Assignment Status (Complete or Incomplete) |
|---|---|---|---|---|
|  |  |  |  |  |
|  |  |  |  |  |
|  |  |  |  |  |
|  |  |  |  |  |
|  |  |  |  |  |
|  |  |  |  |  |
|  |  |  |  |  |
|  |  |  |  |  |
|  |  |  |  |  |
|  |  |  |  |  |

**Vocabulary**

| Term | Description | Example |
|---|---|---|
| 1. Expression versus equation |  |  |
| 2. Coefficient |  |  |
| 3. Constant |  |  |
| 4. Distributive property |  |  |
| 5. Inequality |  |  |
| 6. Solution to an equation versus an inequality |  |  |

| Test or Quiz Name | Testing Date | Original Test Score<br>*Original score you earned when you took the test* | Correct and Reflect Score<br>*3—Completed correctly*<br>*2—Incomplete and needs corrections*<br>*1—Never turned in* | Retake Score<br>*Required if your score is less than 2.5* | Gradebook Score for This Test<br>*Most recent score on the original test or the retake* |
|---|---|---|---|---|---|
| | | | | | |
| | | | | | |
| | | | | | |
| | | | | | |
| | | | | | |

**Check one:**
☐ I am maintaining a proficient level for this class (3.0).   ☐ I need to put forth more effort in order to be proficient (3.0).

**Action steps for improvement:** Place a check mark in the shaded boxes under the tasks you will do to relearn essential standards, as needed.

| Test or Quiz Name | Morning (a.m.) Tutoring With Teacher<br>*Available from 7:45 to 8:15* | Afternoon (p.m.) Tutoring With Teacher<br>*Available from 3:15 to 3:45* | Tutoring With a Peer<br>*Before school, at lunch, or after school* | Completed Unit Study Guide | Homework Status<br>*I have completed all my homework for this test.* | Correct and Reflect<br>*I can do this on my own and finish before future due dates.* | After-School Intervention<br>*I need to sign up for this every weekend.* |
|---|---|---|---|---|---|---|---|
| | | | | | | | |
| | | | | | | | |
| | | | | | | | |
| | | | | | | | |

**How Can I Learn the Target and Raise My Grade?**

Successful mathematics students:
- Come to class prepared (organized binder with completed homework, pencils, calculator, textbook, and paper) and are ready to learn every day
- Ask for help before they take an assessment so there are no surprises
- Ask questions in class, do their very best on every assignment, and know where they still need extra practice
- Complete the correct and reflect in a timely manner if they earn a failing grade on any test (Retakes are available once the correct and reflect is done correctly.)

What do you still need to work on in order to be successful this year?


Everyone can be successful!

Student signature: _____   Parent or guardian signature: _____

*Source: © 2016 by Ridgeview Middle School. Used with permission.*

**Figure 11.15: Example grade 7 student tracker.**

| **Geometry 1: Introduction to Geometry** | | | |
|---|---|---|---|
| Name _____'s Assignment Tracker   Period: _____   Date: _____ | | | |
| Assignments: Complete the table to track progress on independent practice. | | | |
| **Assignments and Notes (HW = Homework)** | **Essential Learning Standard Addressed** | **Due Date** | **Completed** |
| Day 1: Spiral Review HW | Prerequisites | | |
| Day 2: Spiral Review HW | Prerequisites | | |
| Day 3: G-CO.1 HW | HSG-CO.A.1 | | |
| Day 4: G-CO.1 HW Midpoint | HSG-CO.A.1 | | |
| Day 5: City Planning Project | HSG-CO.A.1 | | |
| Days 6–7: Construction Blueprint | HSG-CO.D.12 | | |
| Day 8: Segment and Angle Addition HW | HSG-CO.A.1 | | |
| Days 9–10: Bisector HW | HSG-CO.A.1 | | |
| Day 11: Angle Pair Relationship HW | HSG-CO.A.1 | | |
| Day 12: Angle Pair Relationship HW | HSG-CO.A.1 | | |
| Day 13: HW Unit Reviews | HSG-CO.A.1<br>HSG-CO.D.12 | | |
| Test Review | All Standards | | |

**Check for Understanding:** Self-assess your current understanding of each standard throughout the unit.

| **Essential Learning Standard** | **Self-Assessment During the Unit** | | | **Assessments** | |
|---|---|---|---|---|---|
| | | | | **Unit Review** | **Unit Assessment** |
| HSG-CO.A.1: Know precise definitions of angle, circle, perpendicular line, parallel line, and line segment, based on the undefined notions of point, line, distance along a line, and distance around a circular arc. | | | | | |
| HSG-CO.D.12: Make formal geometric constructions with a variety of tools and methods (compass and straightedge, string, reflective devices, paper folding, dynamic geometric software, etc.). Copying a segment; copying an angle; bisecting a segment; bisecting an angle; constructing perpendicular lines, including the perpendicular bisector of a line segment. | | | | | |

*Source for standards: NGA & CCSSO, 2010.*

**Figure 11.16: Sample high school student tracker.**

## Conclusion

Consider how your team will plan for student investment. Student investment involves students finding meaning and relevance in what they are learning and understanding their progress toward learning essential standards. Students know what they have learned and what they have not learned *yet*. Encourage students to set goals for future learning and monitor their progress throughout each unit.

Investing students in their learning also requires them to reflect using rubrics or success criteria during class and after a common assessment. In Tier 1, your team can provide goal cards, proficiency scales, and rubrics or success criteria to clarify your expectations for student learning. After each common assessment, students reflect to identify what the evidence says about their level of learning. You might also create a survey to hear from students about how they learn best in your class.

| **Unit:** Grade 2 Word Problem Unit | | | **Date:** January 10–14 | |
|---|---|---|---|---|
| **Directions:** After the lesson, circle + if you learned the target and ? if you still have questions. | | | | |
| **Monday** | **Tuesday** | **Wednesday** | **Thursday** | **Friday** |
| I can add to 100 and show my thinking. | I can add to 100 and show my thinking. | I can add to 100 and show my thinking. | I can add to 100 and show my thinking. | I can add to 100 and show my thinking. |
| Add using base ten blocks. 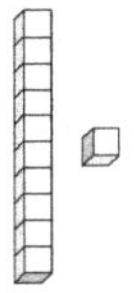 | Add using place value charts. <br><br> \| Tens \| Ones \| | Add numbers without regrouping. <br><br> 43 + 51 = ? | Add numbers with regrouping. <br><br> 27 + 64 = ? | Add numbers with regrouping. <br><br> 35 + 27 = ? |
| +     ? | +     ? | +     ? | +     ? | +     ? |

**Figure 11.17:** Example weekly grade 2 student tracker.

*Visit **go.SolutionTree.com/MathematicsatWork** for a free reproducible version of this figure.*

From their reflections and learning evidence on common assessments, any Tier 2 interventions become opportunities to learn a standard that has not yet been mastered (or to extend learning). Students see the connection from their work in Tier 1 to the learning needed in Tier 2. Most importantly, through additional Tier 2 learning opportunities, students learn that they can now demonstrate learning of the target or standard they had previously not yet learned to proficiency. Through Tier 1 and Tier 2 learning experiences, students see they are learning grade- or course-level standards and, ultimately, improve their self-efficacy, mathematical identity, and agency.

## Questions to Consider for Next Steps

As a collaborative team, use the following questions to reflect on your current practices and determine any next steps related to student investment strategies.

1. How are you and your team currently asking students to set goals and track their learning?

2. Which types of tools (for example, goal cards, rubrics, success criteria, proficiency scales, or surveys) do you and your team want to try as you grow student investment?

3. How will you know when students are developing the skill of setting goals and tracking their learning?

4. What strategies might your team try to grow student investment in Tier 2 interventions and extensions?

5. How can your team connect the student investment tools you are using during Tier 1 with Tier 2 learning opportunities?

# Epilogue

> Don't tell me you "believe all kids can learn";
> tell me what you are doing about the kids who aren't learning.
>
> —Rick DuFour

The strategies in this book are meant to strengthen Tier 1 initial instruction as well as launch new ideas for re-engaging learners in Tier 2 who have not yet learned essential mathematics content in your grade level or course. These strategies are not assigned to Tier 1 or Tier 2 instruction specifically; rather, they can be used in both tiers to grow student learning.

You might use various strategies during initial instruction at the beginning of a unit, as a response to a misconception discovered in a class assignment, or as necessary Tier 2 re-engagement following a common end-of-unit assessment. Depending on your grade level or learning targets, you may even share these strategies with your intervention team to use during Tier 3 instruction.

Working together as a collaborative mathematics team enables you and your colleagues to reach every student. Together, clarify what students will learn unit by unit, create common assessments to gather evidence of that learning, analyze the assessment evidence to reveal any trends in misconceptions or errors students are making, and then create re-engagement learning experiences for students in Tier 1 or Tier 2, as needed. This level of collaboration ensures that every student's needs are met.

As a team, use student work to determine which instructional practices positively affect learning, which need to be adjusted to yield better results, and which need to be abandoned altogether. These team actions result from answering the four critical questions of a PLC at Work (DuFour et al., 2024), which are addressed in more detail in the Mathematics at Work *Every Student Can Learn Mathematics* series. By engaging in this ongoing, cyclical work, you and your team gain clarity about the content you are teaching, add to your bank of strategies that support student learning, and grow your collective teacher efficacy.

While the strategies shared in this book can be used as part of Tier 1 or Tier 2 mathematics learning, collaborative team members ask themselves different planning questions for each tier. Figure E.1 (page 172) provides example questions to consider when planning for initial instruction as prevention (Tier 1) and using common assessment data to plan for team interventions and extensions (Tier 2).

An important team action contributing to student success is using common assessment data to inform instructional decisions, especially for Tier 2 re-engagement. Work together to determine how you and your team will monitor progress unit by unit and throughout the year. Common mid-unit assessments, exit tickets, and end-of-unit assessments provide your team information in the form of student work (written or observed) that can be sorted by proficiency. Ask yourselves, "Which students have mastery? Which students are close? Which students have several misconceptions? Why?" and "Which instructional practices are working?"

| Tier 1 | Tier 2 |
|---|---|
| • What essential standards are students learning in the next unit?<br>• What content have students learned in the last unit, earlier this year, or last year that we ,can connect to the learning in this unit?<br>• How will we intentionally connect prior knowledge to current learning throughout the unit?<br>• What is the mathematical rigor (conceptual understanding, application, and procedural fluency) needed for students to learn the full rigor of the essential standards?<br>• What mathematical tasks will we use in class?<br>• How will we emphasize mathematical language and student-to-student discourse during the lessons?<br>• How will we differentiate learning during lessons (using high-level tasks and small groups)?<br>• How will we gather evidence of student learning during lessons and instructionally respond? | • What common misconceptions or mistakes are students making, as evidenced from their work on common assessments or daily formative assessments?<br>• How will we collectively respond to common misconceptions or mistakes students are making? What is a meaningful extension for students, if needed?<br>• What strategies have students learned that we might want to use again to help them make connections when re-engaging them in learning?<br>• How might we use the concrete, representational, abstract (CRA) framework to re-engage students in learning?<br>• What strategies and models should we use as part of our team intervention?<br>• What conceptual understanding may we need to revisit with students to make connections to the essential standards not yet learned? |

**Figure E.1: Sample questions to answer as a team when planning for Tier 1 and Tier 2 instruction.**

*Visit **go.SolutionTree.com/MathematicsatWork** for a free reproducible version of this figure.*

Sorting student work impacts how you respond as a team or within individual classrooms. Understanding the different minor errors and major misconceptions revealed in student work allows your team to determine which learning targets are necessary for students to re-engage in learning and the most useful strategies to employ to grow student learning to grade or course level and beyond (see appendix A, page 173, for examples of data protocols).

Consider as a team how you are capturing and archiving your learning unit by unit. Your team organizes their work through electronic folders and ensures every team member has access to the products being used to plan student learning. Electronically housing shared resources and tools for responding to student learning needs serves two purposes: (1) you and your colleagues will be able to quickly find and access files, notes, and assessment tools when needed, and (2) you can record and save any artifacts needed to engage in that practice again once your team identifies an instructional practice or routine that is helping students learn at grade or course level. The mindset of *this is how I do it* is no longer acceptable if you have compelling evidence that another strategy or shift in implementation will improve student learning.

Additionally, if you or a colleague engage students in learning through a successful structure, strategy, or practice, share it with your teammates at your grade or course level, and potentially vertically, depending on the scope of the impact. When you enter a school, you do not work in isolation to have *some* students learn. You work in collaborative teams to embrace the promise that learning should be equitable for *all*, including having access to meaningful Tier 1 core instruction and Tier 2 interventions and extensions, as needed. Make a promise to every family that their child did not get the teacher; their child got the team. And that team will work together to ensure every student learns.

This book provides guidance for teachers and teams while planning mathematics instruction to accelerate learning through Tier 1 and Tier 2 learning experiences. Explore the strategies shared, and work with your colleagues to determine which ones contribute to more students learning mathematics in your school and across your team. There is no time to lose. You and your team ensure every student learns essential grade- or course-level mathematics through your intentional planning and targeted interventions. Believe in your students. Believe in yourselves. Your students are counting on you.

# Appendix A

# Data Analysis Protocols

Team-created Tier 1 and Tier 2 targeted interventions and extensions are determined using evidence of student learning shown on common mid-unit and end-of-unit assessments. Bring student work to team meetings to analyze data and see trends in student thinking and reasoning across your team. Student work often reveals the next instructional steps to re-engage students in learning, and it provides an opportunity for you and your team to analyze the effectiveness of your instructional practices.

As shared in chapter 1 (page 9), responding to common mid-unit assessments is often done in Tier 1 as prevention before the unit ends, and responding to common end-of-unit assessments often occurs during Tier 2 interventions and extensions. The data reveal the number of students needing targeted re-engagement, which also informs whether interventions and extensions are done during Tier 1 or Tier 2.

Before analyzing student data, discuss as a team how you are scoring your common assessments and what students must demonstrate to reach proficiency. Calibrate your scoring of assessments to ensure your team is evaluating student learning consistently and equitably. The design of the assessments, calibration, and data analysis are discussed in greater detail in *Mathematics Assessment and Intervention in a PLC at Work, Second Edition* (Schuhl et al., 2024).

In general, your team should sort student work into piles—categories include students who are far from proficient or have a minimal understanding, students who are close to proficient or have a partial understanding, students who are proficient, and, possibly, students who demonstrate an advanced or thorough understanding (not all assessments will have advanced questions or advanced student work for this last pile). After determining how many students are in each pile, identify trends in student thinking for each category, and then create a team-targeted intervention and extension plan for each student group.

Consider the following steps when determining trends in student thinking (including common misconceptions and reasoning errors as well as successful strategies used).

1. Start your team discussion by identifying what the proficient (and advanced) students did that put their work in the proficient (or advanced) pile. For example, did students all use a number line? Did they check their answers? Did they use a particular model or strategy?

2. Next, identify a trend among the students who were close to proficiency or had a partial understanding. What did these students do mathematically? Can you identify targeted interventions that might move these students to proficient?

3. Finally, look at the students with a minimal understanding. There may or may not be a trend. If so, identify the trend; if not, write

*no trend.* As before, identify what the close-to-proficient students could do, and create a targeted intervention to move students with minimal understanding to close to proficient. Your intervention might also be a prior knowledge standard to move students forward in their learning.

4. Finally, create a team-targeted intervention and extension plan for each group of students. Chapter 1 (page 9) shares ideas for shuffling students, creating stations, and other options to implement interventions and extensions across the team.

The following are two possible data protocols from *Mathematics Assessment and Intervention in a PLC at Work, Second Edition* (Schuhl et al., 2024) that your team may want to use to analyze student work on common assessments and determine a Tier 1 or Tier 2 re-engagement plan.

# Student Work Protocol

| Student Work Protocol |
|---|
| Learning targets or essential learning standards: |

1. Analyze the expectation for student work or performance at each level. What are the criteria to assess this work?

| Beginning Standard | Approaching Standard | Meeting Standard | Exceeding Standard |
|---|---|---|---|
| | | | |

2. Sort student work, and list student names at the appropriate level or the number of students by teacher in each level (with student names identified in the gradebook). Identify what percentage of students are at each level across the team.

| Beginning Standard | Approaching Standard | Meeting Standard | Exceeding Standard |
|---|---|---|---|
| Percent of team: | Percent of team: | Percent of team: | Percent of team: |

3. Choose a couple of student samples from each level, and describe the evidence of student learning. Focus on the following four areas. Document the trends of student thinking in each level.
   a. Demonstrates deep conceptual understanding
   b. Shows procedural knowledge of mathematical content
   c. Demonstrates skills and understanding in problem solving
   d. Demonstrates effective communication

| Beginning Standard | Approaching Standard | Meeting Standard | Exceeding Standard |
|---|---|---|---|
| | | | |

4. Determine a plan for next instructional steps as a team. What is the targeted and specific plan for each group of students to re-engage students in learning (Tier 1 or Tier 2 interventions)? Identify any future changes in instruction on your team unit plan.

| Beginning Standard | Approaching Standard | Meeting Standard | Exceeding Standard |
|---|---|---|---|
| | | | |

*Source: Schuhl, S., Kanold, T. D., Toncheff, M., Barnes, B., Kanold-McIntyre, J., Larson, M. R., et al. (2024). Mathematics assessment and intervention in a PLC at Work (2nd ed.). Bloomington, IN: Solution Tree Press, p. 84.*

# Essential Learning Standard Analysis Protocol: Grade 4 Sample

## Essential Learning Standard Analysis Protocol

1. Define each essential standard on the assessment and describe the expectation of proficiency. Write your definitions and descriptions in the following chart.

| | Essential Standard 1 | Essential Standard 2 | Essential Standard 3 | Essential Standard 4 |
|---|---|---|---|---|
| Expectations of Proficiency | I can explain why fractions are equivalent and create equivalent fractions. Student creates two equivalent fractions from a given fraction or model and explains why the fractions are equivalent. | I can compare two fractions and explain my thinking. Student compares two fractions using <, >, or = when given a model and without a model and shows an understanding that fractions can only be compared when they reference the same whole. | I can add and subtract fractions and show my thinking. Student adds and subtracts fractions with like denominators, decomposes a fraction more than one way, and shows thinking using models and equations. | I can multiply a fraction by a whole number and explain my thinking. Student multiplies a fraction by a whole number using models and equations to show thinking. |

2. Determine the number and percentage of students proficient or advanced on the assessment for each standard by teacher, and then determine the total number of proficient or advanced students within the team. Write the information in the following chart.

| | Essential Standard 1 | | Essential Standard 2 | | Essential Standard 3 | | Essential Standard 4 | | Total Number of Students |
|---|---|---|---|---|---|---|---|---|---|
| | Number | Percent | Number | Percent | Number | Percent | Number | Percent | |
| Teacher A | 20 | 65% | 10 | 32% | 22 | 71% | 28 | 90% | 31 |
| Teacher B | 19 | 68% | 12 | 43% | 20 | 71% | 26 | 93% | 28 |
| Teacher C | 20 | 65% | 8 | 29% | 28 | 90% | 25 | 81% | 31 |
| Total Team | 59 | 66% | 30 | 33% | 70 | 77% | 79 | 88% | 90 |

3. For each standard, determine which students have unsatisfactory knowledge, which have limited knowledge, which are proficient, and which are advanced by teacher and as a team.

### Essential Standard 1

| | Beginning | Approaching | Proficient | Advanced | Total Number of Students |
|---|---|---|---|---|---|
| Teacher A | 2 | 9 | 10 | 10 | 31 |
| Teacher B | 8 | 1 | 19 | 0 | 28 |
| Teacher C | 11 | 0 | 16 | 4 | 31 |
| Total Team | 21 | 10 | 45 | 14 | 90 |

### Essential Standard 2

| | Beginning | Approaching | Proficient | Advanced | Total Number of Students |
|---|---|---|---|---|---|
| Teacher A | 10 | 11 | 2 | 8 | 31 |
| Teacher B | 10 | 6 | 12 | 0 | 28 |
| Teacher C | 19 | 4 | 4 | 4 | 31 |
| Total Team | 39 | 21 | 18 | 12 | 90 |

| Essential Standard 3 | Beginning | Approaching | Proficient | Advanced | Total Number of Students |
|---|---|---|---|---|---|
| Teacher A | 4 | 5 | 7 | 15 | 31 |
| Teacher B | 7 | 1 | 19 | 1 | 28 |
| Teacher C | 3 | 0 | 18 | 10 | 31 |
| **Total Team** | 14 | 6 | 44 | 26 | 90 |

| Essential Standard 4 | Beginning | Approaching | Proficient | Advanced | Total Number of Students |
|---|---|---|---|---|---|
| Teacher A | 1 | 2 | 20 | 8 | 31 |
| Teacher B | 0 | 2 | 22 | 4 | 28 |
| Teacher C | 6 | 0 | 24 | 1 | 31 |
| **Total Team** | 7 | 4 | 66 | 13 | 90 |

1. Which essential standards were student strengths? What instructional strategies impacted student thinking?

    Our students are doing well with adding and subtracting fractions with common denominators and multiplying fractions by a whole number. Having the students engage in number talks during this unit has really helped students make connections between the models that students create and develop the thought processes needed.

2. In which areas did individual teachers' students struggle? In which areas did our team's students struggle? What is the cause? How will we respond?

    Only 66% of our students are proficient with equivalent fractions. Teacher A has been using more manipulatives and will use them with students we define need intervention from all three classrooms.

    As a team, students are struggling with comparing two fractions when the denominator is not common. They are confusing the models or are not precise when using a circle to compare fractions. We will create a plan for the students who need more time and support to include a focus on fraction representation using the rectangular model and the applet from NCTM.

3. Which students need additional time and support to learn the standards? What is our plan?

    Next week during intervention time, we will use the following schedule:

    Monday and Tuesday—Teacher A and support staff will work with the 39 identified students on comparing two fractions. Teachers B and C will use the recipe task to stretch students' understanding of uncommon denominators.

    We will also use small-group instruction and centers during the week to do more work with manipulatives. Teacher A is going to bring her manipulatives to share at the next team meeting.

4. Which students need extension or enrichment? What is our plan?

    See notes on the use of the recipe task.

*Source: Schuhl, S., Kanold, T. D., Toncheff, M., Barnes, B., Kanold-McIntyre, J., Larson, M. R., et al. (2024). Mathematics assessment and intervention in a PLC at Work (2nd ed.). Bloomington, IN: Solution Tree Press, pp. 87–88.*

# Appendix B

# Cognitive-Demand-Level Task Analysis Guide

# Cognitive-Demand Levels of Mathematical Tasks

**Directions:** This appendix provides criteria you can use to evaluate the cognitive-demand level of the tasks you choose each day. As you develop your assessments, unit plans, and lessons, use this tool to ensure a balance of low-level and high-level tasks for students as they learn the standards.

| Lower-Level Cognitive Demand | Higher-Level Cognitive Demand |
| --- | --- |
| **Memorization Tasks**<br>• These tasks involve reproducing previously learned facts, rules, formulae, or definitions to memory.<br>• They cannot be solved using procedures because a procedure does not exist or because the time frame in which the task is being completed is too short to use the procedure.<br>• They are not ambiguous; such tasks involve exact reproduction of previously seen material and what is to be reproduced is clearly and directly stated.<br>• They have no connection to the concepts or meaning that underlie the facts, rules, formulae, or definitions being learned or reproduced. | **Procedures With Connections Tasks**<br>• These procedures focus students' attention on the use of procedures for the purpose of developing deeper levels of understanding of mathematical concepts and ideas.<br>• They suggest pathways to follow (explicitly or implicitly) that are broad general procedures that have close connections to underlying conceptual ideas as opposed to narrow algorithms that are opaque with respect to underlying concepts.<br>• They usually are represented in multiple ways (for example, visual diagrams, manipulatives, symbols, or problem situations). They require some degree of cognitive effort. Although general procedures may be followed, they cannot be followed mindlessly. Students need to engage with the conceptual ideas that underlie the procedures in order to successfully complete the task and develop understanding. |
| **Procedures Without Connections Tasks**<br>• These procedures are algorithmic. Use of the procedure is either specifically called for, or its use is evident based on prior instruction, experience, or placement of the task.<br>• They require limited cognitive demand for successful completion. There is little ambiguity about what needs to be done and how to do it.<br>• They have no connection to the concepts or meaning that underlie the procedure being used.<br>• They are focused on producing correct answers rather than developing mathematical understanding.<br>• They require no explanations or have explanations that focus solely on describing the procedure used. | **Doing Mathematics Tasks**<br>• Doing mathematics tasks requires complex and nonalgorithmic thinking (for example, the task, instructions, or examples do not explicitly suggest a predictable, well-rehearsed approach or pathway).<br>• It requires students to explore and understand the nature of mathematical concepts, processes, or relationships.<br>• It demands self-monitoring or self-regulation of one's own cognitive processes.<br>• It requires students to access relevant knowledge and experiences and make appropriate use of them in working through the task.<br>• It requires students to analyze the task and actively examine task constraints that may limit possible solution strategies and solutions.<br>• It requires considerable cognitive effort and may involve some level of anxiety for the student due to the unpredictable nature of the required solution process. |

*Source: Smith, M. S., & Stein, M. K. (1998). Selecting and creating mathematical tasks: From research to practice. Mathematics Teaching in the Middle School, 3(5), 344–350, p. 348.*

# Appendix C

# Team Actions to Avoid and Consider for Tier 1 and Tier 2

# Team Actions to Avoid and Consider for Tier 1 and Tier 2

| Actions to Avoid | Actions to Consider |
| --- | --- |
| **Tier 1**<br>• Teach page by page in the book through each unit, and get as far as you can.<br>• Reteach everything students should have learned before starting grade- or course-level content.<br>• Move students to a lower grade or course, or retain students.<br>• Create ability groups, and teach each group separately with different learning expectations.<br>• Talk louder and slower.<br>• Only give steps and have students practice.<br>• Use low-level tasks and skill-based problems only.<br>• Slow instruction down for all students.<br>• Teach everything from each lesson in the book.<br>• Have students learn through memorization rather than understanding. | **Tier 1**<br>• Clarify the guaranteed and viable curriculum all students will learn (DuFour et al., 2024).<br>• Teach prerequisite skills students need to learn within appropriate units, always concluding with grade- or course-level standards (Schuhl et al., 2020).<br>• Create and give common assessments for formative learning (DuFour et al., 2024).<br>• Have students learn in mixed-ability groups of four with intentional tasks during each lesson (Toncheff et al., 2024).<br>• Engage students in discourse so they learn from one another (Hattie, 2023).<br>• Focus on students learning through conceptual understanding (NCTM, 2014).<br>• Use a balance of high- and low-level cognitive-demand tasks (include word problems; Smith & Stein, 2011).<br>• Make connections to each lesson; build any needed prior knowledge that is missing (Toncheff et al., 2024).<br>• Emphasize NCTM's (2014) productive beliefs. |
| **Tier 2**<br>• Send students out to other adults on campus to be "fixed."<br>• Solely use a computer program that determines which standards a student has learned and creates a learning plan independent of the learning happening in class.<br>• Group students for intervention based on an overall test percentage (for example, 65 percent), when the score does not detail by target what a student has learned or has not learned yet.<br>• Teach standards the same way they were taught during Tier 1 core instruction.<br>• Focus solely on basic facts and algorithms.<br>• Only use interim assessment data or progress-monitoring data to determine placement, and leave students in the same intervention for extended periods of time. | **Tier 2**<br>• Use data from common assessments to target interventions using student misconceptions (Mattos et al., 2025).<br>• Address prior knowledge needed for the current or next unit (preteach; Schuhl et al., 2020).<br>• Use word problems to develop context for learning mathematics standards (NCTM, 2014).<br>• Use the concrete-representational-abstract (CRA) model for learning: What concrete models or manipulatives can students use to make sense of mathematics (C)? What representations (drawings) can students use to see mathematics (R)? How can students use numbers and symbols to abstractly document the mathematics (A)? (Flores, Hinton, & Burton, 2016).<br>• Focus on number sense and an understanding of mathematics over memorized facts (NCTM, 2014).<br>• Review the eight recommendations from the Institute of Education Sciences (Gersten et al., 2009). |

*Source: Schuhl, S. (2021). Working together to ensure all students learn mathematics. In S. V. Kramer (Ed.),* Charting the course for collaborative teams: Lessons from priority schools in a PLC at Work® *(pp. 131–147). Bloomington, IN: Solution Tree Press, p. 135.*

# References and Resources

Achieve the Core. (n.d.). *Coherence map.* Accessed at https://tools.achievethecore.org/coherence-map on December 21, 2023.

Ainsworth, L. (2004). *Power standards: Identifying the standards that matter most.* Englewood, CO: Advanced Learning Press.

Ainsworth, L. (2013). *Prioritizing the common core: Identifying specific standards to emphasize the most.* Englewood, CO: Lead and Learn Press.

Ainsworth, L., & Viegut, D. (2006). *Common formative assessments: How to connect standards-based instruction and assessment.* Thousand Oaks, CA: Corwin Press.

Ainsworth, L., & Viegut, D. (2015). *Common formative assessments 2.0: How teacher teams intentionally align standards, instruction, and assessment.* Thousand Oaks, CA: Corwin Press.

Barbieri, C. A., Miller-Cotto, D., Clerjuste, S. N., & Chawla, K. (2023). A meta-analysis of the worked examples effect on mathematics performance. *Educational Psychology Review, 35*(1), Article 11. https://doi.org/10.1007/s10648-023-09745-1.

Barth, R. S. (1990). *Improving schools from within: Teachers, parents, and principals can make the difference.* San Francisco: Jossey-Bass.

Battelle for Kids. (2013). *Vertical progression guide for the Common Core: Mathematics K–12.* Columbus, OH: Author.

Battelle for Kids. (2019). *Framework for 21st century learning: A unified vision for learning to ensure student success in a world where change is constant and learning never stops.* Accessed at www.battelleforkids.org/wp-content/uploads/2023/11/P21_Framework_Brief.pdf on June 27, 2024.

Bay-Williams, J. M., & Kling, G. (2014). Enriching addition and subtraction fact mastery through games. *Teaching Children Mathematics, 21*(4), 238–247.

Berlin, R., & Cohen, J. (2020). The convergence of emotionally supportive learning environments and college and career ready mathematical engagement in upper elementary classrooms. *AERA Open, 6*(3), 1–20. https://doi.org/10.1177/2332858420957612.

Black, P., & Wiliam, D. (1998). Inside the black box: Raising standards through classroom assessment. *Phi Delta Kappan, 80*, 139–148.

Boaler, J. (2016). *Mathematical mindsets: Unleashing students' potential through creative math, inspiring messages and innovative teaching.* San Francisco: Jossey-Bass.

BrainyQuote. (n.d.). *Bill Gates quotes.* Accessed at www.brainyquote.com/quotes/bill_gates_626252?src=t_feedback on November 19, 2023.

Brown, T., & Ferriter, W. M. (2021). *You can learn! Building student ownership, motivation, and efficacy with the PLC at Work process.* Bloomington, IN: Solution Tree Press.

Bruner, J. S., & Kenney, H. J. (1965). Representation and mathematics learning. *Monographs of the Society for Research in Child Development, 30*(1), 50–59.

Burns, M. (2007). *About teaching mathematics: A K–8 resource* (3rd ed.). Sausalito, CA: Math Solutions.

California Department of Education. (2023). *Mathematics framework, chapter 1: Mathematics for all: Purpose, understanding, and connection.* Accessed at www.cde.ca.gov/ci/ma/cf on December 6, 2023.

Callahan, S. (2012). *Effect of prior grade-level content on pretest reliability* [Research brief]. Scottsdale, AZ: Assessment Technology Incorporated. Accessed at www.ati-online.com/pdfs/researchK12/EffectofPriorGradeLevel ContentonPretestReliability.pdf on March 15, 2024.

Casa, T. M., Firmender, J. M., Cahill, J., Cardetti, F., Choppin, J. M., Cohen, J., et al. (2016). *Types of and purposes for elementary mathematical writing: Task force recommendations.* Mansfield, CT: University of Connecticut Neag School of Education. Accessed at https://mathwriting.education.uconn.edu/wp-content/uploads/sites/1454/2021/09/Elementary-Mathematical-Writing-White-Paper.pdf on March 15, 2024.

Casa, T. M., Gilson, C. M., Bruce-Davis, M. N., Gubbins, E. J., Hayden, S. M., & Canavan, E. J. (2022). Maximizing the potential of mathematical writing prompts. *Mathematics Teacher: Learning and Teaching PK–12, 115*(8), 538–550. https://doi.org/10.5951/MTLT.2021.0287.

Center for Teaching and Learning. (n.d.). *Types of rubrics.* Accessed at https://resources.depaul.edu/teaching-commons/teaching-guides/feedback-grading/rubrics/Pages/types-of-rubrics.aspx on January 20, 2024.

Chappuis, J., & Stiggins, R. (2020). *Classroom assessment for student learning: Doing it right—Using it well* (3rd ed.). New York: Pearson.

Chenoweth, K. (2021). *Districts that succeed: Breaking the correlation between race, poverty, and achievement.* Cambridge, MA: Harvard Education Press.

Common Core Standards Writing Team. (2011, May 29). *Progressions for the Common Core State Standards in mathematics (draft).* Washington, DC: Author. Accessed at https://achievethecore.org/content/upload/Draft-K-5%20Progression%20on%20Counting%20and%20Cardinality%20and%20Operations%20and%20Algebraic%20Thinking.pdf on January 15, 2024.

Common Core Standards Writing Team. (2013, July 2). *Progressions for the Common Core State Standards in mathematics (draft).* Tucson, AZ: University of Arizona. Accessed at https://achievethecore.org/page/254/progressions-documents-for-the-common-core-state-standards-for-mathematics on May 17, 2024.

Dale, E., & O'Rourke, J. (1986). *Vocabulary building: A process approach.* Columbus, OH: Zaner-Bloser.

Danielson, C. (2019). *Which one doesn't belong? Playing with shapes.* Watertown, MA: Charlesbridge.

Darling-Hammond, L., Flook, L., Cook-Harvey, C., Barron, B., & Osher, D. (2020). Implications for educational practice of the science of learning and development. *Applied Developmental Science, 24*(2), 97–140. https://doi.org/10.1080/10888691.2018.1537791.

Darling-Hammond, L., Hyler, M. E., & Gardner, M. (2017, June 5). *Effective teacher professional development.* Palo Alto, CA: Learning Policy Institute. Accessed at https://learningpolicyinstitute.org/product/teacher-prof-dev on October 18, 2023.

Dewey, J. (1933). *How we think: A restatement of the relation of reflective thinking to the educative process.* Boston: D.C. Heath.

Dimich, N. (2014). *Design in five: Essential phases to create engaging assessment practice.* Bloomington, IN: Solution Tree Press.

Dimich, N. (2024). *Design in five: Essential phases to create engaging assessment practice* (2nd ed.). Bloomington, IN: Solution Tree Press.

Dixon, J. K., Nolan, E. C., Lott Adams, T., Brooks, L. A., & Howse, T. D. (2016). *Making sense of mathematics for teaching grades K–2.* Bloomington, IN: Solution Tree Press.

DuFour, R., DuFour, R., Eaker, R., Many, T. W., Mattos, M., & Muhammad, A. (2024). *Learning by doing: A handbook for Professional Learning Communities at Work* (4th ed.). Bloomington, IN: Solution Tree Press.

Dweck, C. (2015, September 22). Carol Dweck revisits the 'growth mindset' [Blog post]. *Ed Week*. Accessed at www.edweek.org/leadership/opinion-carol-dweck-revisits-the-growth-mindset/2015/09 on March 18, 2024.

Erkens, C., Schimmer, T., & Dimich, N. (2017). *Essential assessment: Six tenets for bringing hope, efficacy, and achievement to the classroom*. Bloomington, IN: Solution Tree Press.

Flores, M. M., Hinton, V. M., & Burton, M. E. (2016). Teaching problem solving to students receiving tiered interventions using the concrete-representational-abstract sequence and schema-based instruction. *Preventing School Failure, 60*(4), 345–355.

Frayer, D. A., Fredrick, W. C., & Klausmeier, H. J. (1969). *A schema for testing the level of concept mastery* (Working Paper No. 16). Madison: Wisconsin Research and Development Center for Cognitive Learning.

Fuchs, L. S., Newman-Gonchar, R., Schumacher, R., Dougherty, B., Bucka, N., Karp, K. S., et al. (2021, March). *Assisting students struggling with mathematics: Intervention in the elementary grades*. Washington, DC: National Center for Education Evaluation and Regional Assistance. Accessed at https://ies.ed.gov/ncee/WWC/Docs/PracticeGuide/WWC2021006-Math-PG.pdf on March 18, 2024.

Gallagher, M. A., Ellis, L., & Weiland, T. (2021). Making word problems meaningful. *Mathematics Teacher: Learning and Teaching PK–12, 114*(8), 580–590.

Garrett, C., Ritchie, S., & Phillips, E. (2022, September). *Children come first: The one doing the talking is the one doing the learning*. Browns Summit, NC: Region 6 Comprehensive Center. Accessed at https://region6cc.uncg.edu/wp-content/uploads/2022/10/TheOneDoingtheTalkingistheOneDoingtheLearning-_2022_RC6_006.pdf on March 18, 2024.

Gersten, R., Beckmann, S., Clarke, B., Foegen, A., Marsh, L., Star, J. R., et al. (2009). *Assisting students struggling with mathematics: Response to Intervention (RtI) for elementary and middle schools (NCEE 2009-4060)*. Washington, DC: National Center for Education Evaluation and Regional Assistance, Institute of Education Sciences, U.S. Department of Education. Accessed at http://ies.ed.gov/ncee/wwc/publications/practiceguides on September 23, 2024.

Gojak, L. M. (2013, January 8). *The power of a good mistake*. Accessed at www.nctm.org/News-and-Calendar/Messages-from-the-President/Archive/Linda-M_-Gojak/The-Power-of-a-Good-Mistake on March 18, 2024.

Gonzalez, J. (2014, May 1). Know your terms: Holistic, analytic, and single-point rubrics [Blog post]. *Cult of Pedagogy*. Accessed at www.cultofpedagogy.com/holistic-analytic-single-point-rubrics on January 20, 2024.

Graham, S., Kiuhara, S. A., & MacKay, M. (2020). The effects of writing on learning in science, social studies, and mathematics: A meta-analysis. *Review of Educational Research, 90*(2), 179–226. https://doi.org/10.3102/0034654320914744.

Gray, E. M., & Tall, D. O. (1994). Duality, ambiguity, and flexibility: A "proceptual" view of simple arithmetic. *Journal for Research in Mathematics Education, 25*(2), 116-140.

Guskey, T. R. (2015). *On your mark: Challenging the conventions of grading and reporting*. Bloomington, IN: Solution Tree Press.

Hamre, B. K., & Pianta, R. C. (2005). Can instructional and emotional support in the first-grade classroom make a difference for children at risk of school failure? *Child Development, 76*(5), 949–967. https://doi.org/10.1111/j.1467-8624.2005.00889.x.

Hattie, J. (2012). *Visible learning for teachers: Maximizing impact on learning*. New York: Routledge.

Hattie, J. (2023). *Visible learning: The sequel: A synthesis of over 2,100 meta-analyses relating to achievement*. New York: Routledge.

Hattie, J., Fisher, D., & Frey, N. (2017). *Visible learning for mathematics, grades K–12: What works best to optimize student learning*. Thousand Oaks, CA: Corwin Mathematics.

Hattie, J., & Timperley, H. (2007). The power of feedback. *Review of Educational Research, 77*(1), 81–112.

Heimberg, J. (2007). *Would you rather?* [Book series]. Sacramento, CA: Seven Footer Press.

Heimberg, J., & Gomberg, D. (n.d.). *Would you rather. . .?* Accessed at www.justinheimberg.com/would-you-rather on June 30, 2024.

Hierck, T. (2017). *Seven keys to a positive learning environment in your classroom*. Bloomington, IN: Solution Tree Press.

Hoegh, J. K. (2020). *A handbook for developing and using proficiency scales in the classroom*. Bloomington, IN: Marzano Resources.

Huinker, D., & Bill, V. (2017). *Taking action: Implementing effective mathematics teaching practices, K–grade 5.* Reston, VA: National Council of Teachers of Mathematics.

Hunter, C. (n.d.). *Which one doesn't belong?* Accessed at https://wodb.ca on March 27, 2024.

Illustrative Mathematics. (n.d.). *Marta's multiplication error.* Accessed at https://tasks.illustrativemathematics.org/content-standards/tasks/1524 on July 2, 2024.

Institute of Education Sciences. (2022, May). *Report on the condition of education 2022.* Washington, DC: National Center for Education Statistics. Accessed at https://nces.ed.gov/pubs2022/2022144.pdf on September 29, 2023.

Joyner, J. M., & Bright, G. W. (2016). *INFORMative assessment: Formative assessment practices to improve mathematics achievement, middle and high school.* Sausalito, CA: Math Solutions.

Karp, K. S., Bush, S. B., & Dougherty, B. J. (2014). Thirteen rules that expire. *Teaching Children Mathematics, 21*(1), 18–25.

Karp, K. S., Dougherty, B. J., & Bush, S. B. (2021). *The math pact, elementary: Achieving instructional coherence within and across grades.* Thousand Oaks, CA: Corwin Press.

Kilpatrick, J., Swafford, J., & Findell, B. (Eds.). (2001). *Adding it up: Helping children learn mathematics.* Washington, DC: National Academies Press. Accessed at https://nap.nationalacademies.org/read/9822/chapter/1 on July 8, 2024.

Klein, A. (2015, April 10). *No Child Left Behind: An overview.* Accessed at www.edweek.org/policy-politics/no-child-left-behind-an-overview/2015/04 on May 17, 2024.

Klein, A. (2022, October 19). *Why the Gates Foundation is investing $1.1 billion in math education.* Accessed at www.edweek.org/teaching-learning/why-the-gates-foundation-is-investing-1-1-billion-in-math-education/2022/10 on March 18, 2024.

Kobett, B. M., & Karp, K. S. (2020). *Strengths-based teaching and learning in mathematics: Five teaching turnarounds for grades K–6.* Thousand Oaks, CA: Corwin Press.

Kostos, K., & Shin, E. (2010). Using math journals to enhance second graders' communication of mathematical thinking. *Early Childhood Education Journal, 38*(3), 223–231. https://doi.org/10.1007/s10643-010-0390-4

Kramer, S. V., & Schuhl, S. (2017). *School improvement for all: A how-to guide for doing the right work.* Bloomington, IN: Solution Tree Press.

Kramer, S. V., & Schuhl, S. (2023). *Acceleration for all: A how-to guide for overcoming learning gaps.* Bloomington, IN: Solution Tree Press.

Lang-Raad, N. D., & Marzano, R. J. (2019). *The new art and science of teaching mathematics.* Bloomington, IN: Solution Tree Press.

Lemon, T., & Hendrickson, S. (2023). Building coherence and progression on sound frameworks. *Mathematics Teacher: Learning and Teaching PK–12, 116*(7), 490–502.

Leonhardt, D. (2023, September 5). *Where are the students?* Accessed at www.nytimes.com/2023/09/05/briefing/covid-school-absence.html on June 14, 2024.

Liljedahl, P. (2021). *Building thinking classrooms in mathematics, grades K–12: 14 teaching practices for enhancing learning.* Thousand Oaks, CA: Corwin Mathematics.

Lough, C. (2020, April 28). *Dylan Wiliam: "Immoral" to teach "too full" curriculum.* Accessed at www.tes.com/news/dylan-wiliam-immoral-teach-too-full-curriculum on April 30, 2020.

Many, T. W., Maffoni, M. J., Sparks, S. K., & Thomas, T. F. (2022). *Energize your teams: Powerful tools for coaching collaborative teams in PLCs at Work.* Bloomington, IN: Solution Tree Press.

Marcus, J. (2023, September 27). *Americans have poor math skills. It's a threat to U.S. standing in the global economy, employers say.* Accessed at https://apnews.com/article/math-scores-china-security-b60b740c480270d552d750c15ed287b6 on January 2, 2024.

Marzano, R. J. (2003). *What works in schools: Translating research into action.* Arlington, VA: ASCD.

Marzano, R. J. (2006). *Classroom assessment and grading that work.* Arlington, VA: ASCD.

Marzano, R. J. (2010). *Formative assessment and standards-based grading.* Bloomington, IN: Marzano Resources.

Marzano, R. J. (2017). *The new art and science of teaching.* Bloomington, IN: Solution Tree Press.

Marzano, R. J. (2023). *Step into student goal setting: A path to growth, motivation, and achievement.* Bloomington, IN: Solution Tree Press.

Marzano, R. J., Warrick, P. B., Rains, C. L., & DuFour, R. (2018). *Leading a high reliability school.* Bloomington, IN: Solution Tree Press.

Maslow, A. H. (1943). A theory of human motivation. *Psychological Review, 50*(4), 370–396. https://doi.org/10.1037/h0054346.

Math is Fun. (n.d.). *Zoomable number line.* Accessed at www.mathsisfun.com/numbers/number-line-zoom.html on March 19, 2024.

Math Learning Center. (n.d.). *Number line.* Accessed at https://apps.mathlearningcenter.org/number-line on March 15, 2024.

Mathematics Education Cooperative. (2024). *Number talks.* Accessed at www.mec-math.org/number-talks on August 5, 2024.

Mattos, M. (2023, April). *Building the pyramid: How to create a highly effective, multitiered system of supports* [Conference presentation]. RTI Institute, Green Bay, Wisconsin.

Mattos, M., Buffum, A., Malone, J., Cruz, L. F., Dimich, N., & Schuhl, S. (2025). *Taking action: A handbook for RTI at Work™* (2nd ed.). Bloomington, IN: Solution Tree Press.

McAnelly, N. (2021, December 9). *How math journals help students process their learning.* Accessed at www.edutopia.org/article/how-math-journals-help-students-process-their-learning on January 15, 2024.

Michigan State University College of Natural Science. (n.d.). *Preparing the classroom environment.* Accessed at https://connectedmath.msu.edu/classroom/getting-organized/classroom-environment.aspx on March 18, 2024.

Miller, S. P., & Hudson, P. J. (2007). Using evidence-based practices to build mathematics competence related to conceptual, procedural, and declarative knowledge. *Learning Disabilities Research and Practice, 22*(1), 47–57.

Moeller, B. (n.d.). Reframing our goals for mathematics education: The importance of nurturing a sense of belonging [Blog post]. *Math for All.* Accessed at https://mathforall.edc.org/nurturing-a-sense-of-belonging on January 15, 2024.

National Center for Education Statistics. (2023). *English learners in public schools.* Accessed at https://nces.ed.gov/programs/coe/indicator/cgf on January 17, 2024.

National Council for Curriculum and Assessment. (2022). *Primary mathematics curriculum draft specification junior infants to sixth class.* Dublin, Ireland: Author. Accessed at https://ncca.ie/media/3148/primary_mathsspec_en.pdf on June 2, 2024.

National Council of Supervisors of Mathematics. (n.d.). *Supporting all students through flexible grouping practices: A position statement from NCSM: Leadership in mathematics education.* Englewood, CO: Author. Accessed at www.mathedleadership.org/wp-content/uploads/2023/10/49085_Flex-Grouping-Position-Paper_digital-version.pdf on March 27, 2024.

National Council of Supervisors of Mathematics. (2019). *NCSM Essential Actions: Instructional leadership in mathematics education.* Englewood, CO: Author.

National Council of Supervisors of Mathematics. (2022). *NCSM Essential Actions: Culturally relevant leadership in mathematics education.* Englewood, CO: Author.

National Council of Supervisors of Mathematics & TODOS. (2021). *Positioning multilingual learners for success in mathematics.* Accessed at www.mathedleadership.org/wp-content/uploads/2021/10/NCSM-TODOS-Multilingual-Learners-Position-Paper-2021_UpdatedLogos.pdf on March 27, 2024.

National Council of Teachers of Mathematics. (1989). *Curriculum and evaluation standards for school mathematics.* Reston, VA: Author.

National Council of Teachers of Mathematics. (2000). *Principles and standards for school mathematics.* Reston, VA: Author.

National Council of Teachers of Mathematics. (2006). *Curriculum focal points for prekindergarten through grade 8 mathematics: A quest for coherence.* Reston, VA: Author.

National Council of Teachers of Mathematics. (2014). *Principles to actions: Ensuring mathematical success for all*. Reston, VA: Author.

National Council of Teachers of Mathematics. (2020). *Catalyzing change in early childhood and elementary mathematics: Initiating critical conversations*. Reston, VA: Author.

National Council of Teachers of Mathematics. (2023, January). *Procedural fluency in mathematics: Reasoning and decision-making, not rote application of procedures position*. Reston, VA: Author. Accessed at www.nctm.org/Standards-and-Positions/Position-Statements/Procedural-Fluency-in-Mathematics on January 4, 2023.

National Council of Teachers of Mathematics & National Council of Supervisors of Mathematics. (2020, June). *Moving forward: Mathematics learning in the era of COVID-19*. Reston, VA: Authors. Accessed at https://nctm.org/uploaded Files/Research_and_Advocacy/NCTM_NCSM_Moving_Forward.pdf on May 23, 2022.

National Council of Teachers of Mathematics, National Council of Supervisors of Mathematics, & Association of State Supervisors of Mathematics. (2021, July). *Continuing the journey: Mathematics learning 2021 and beyond*. Reston, VA: Authors. Accessed at www.nctm.org/uploadedFiles/Research_and_Advocacy/collections/Continuing_the_Journey/NCTM_NCSM_Continuing_the_Journey_Report-Fnl2.pdf on December 5, 2023.

National Governors Association Center for Best Practices & Council of Chief State School Officers. (2010). *Common Core State Standards for mathematics*. Washington, DC: Authors. Accessed at https://ccsso.org/sites/default/files/2017-10/MathStandards50805232017.pdf on December 5, 2023.

Nation's Report Card. (n.d.). *NAEP report card: 2022 NAEP mathematics assessment*. Accessed at www.nations reportcard.gov/highlights/mathematics/2022 on December 15, 2023.

New Teacher Project. (2021, May 23). *Accelerate, don't remediate: New evidence from elementary math classrooms*. Accessed at https://tntp.org/publication/accelerate-dont-remediate on March 19, 2024.

New Teacher Project. (2022, March). *Stronger and clearer each time*. New York: Author. Accessed at https://tntp.org/assets/set-resources/MLL/Stronger_and_Clearer_Each_Time_Strategy.pdf on January 16, 2024.

Nordengren, C. (2023). *Step into student goal setting: A path to growth, motivation, and achievement*. Thousand Oaks, CA: Corwin Press.

Norris, K., & Schuhl, S. (2016). *Engage in the mathematical practices: Strategies to build numeracy and literacy with K–5 learners*. Bloomington, IN: Solution Tree Press.

Ochsendorf, R., & Pyke, C. (2007, April 15–18). *Assessment practices in science curriculum materials research: Do students learn from the pretest?* [Conference paper presentation]. National Association for Research in Science Teaching, New Orleans, Louisiana.

Osterman, K. F. (2023). Teacher practice and students' sense of belonging. In T. Lovat, R. Toomey, N. Clement, & K. Dally (Eds.), *Second international research handbook on values education and student wellbeing* (pp. 971–993). New York: Springer.

Parrish, S. (2010). *Number talks: Helping children build mental math and computation strategies, grades K–5*. Sausalito, CA: Math Solutions.

Parrish, S., & Dominick, A. (2016). *Number talks: Fractions, decimals, and percentages*. Sausalito, CA: Math Solutions.

Parrish, S., & Humphreys, C. (2014). *Teaching mathematics with meaning: Connecting the standards with the classroom*. Reston, VA: National Council of Teachers of Mathematics.

Partnership for 21st Century Skills. (2009). *P21 framework definitions*. Accessed at https://files.eric.ed.gov/fulltext/ED519462.pdf on December 18, 2023.

Partnership for Assessment of Readiness for College and Careers. (2014). *PARCC Model Content Frameworks. Mathematics grades 3–11. Version 4.0*. Accessed at https://dam.assets.ohio.gov/image/upload/transfercredit.ohio.gov/files/math/PARCC_MCF_Mathematics-12-11-2014.pdf on June 2, 2024.

PBS LearningMedia. (n.d.). *PBS TeacherLine tips for developing mathematical thinking*. Accessed at https://mpb.pbslearningmedia.org/resource/ea6a2ce8-5e15-4efa-af55-5685971ea0d4/ea6a2ce8-5e15-4efa-af55-5685971ea0d4 on December 27, 2023.

Reeves, D. B. (2002). *The leader's guide to standards: A blueprint for educational equity and excellence*. San Francisco: Jossey-Bass.

Reeves, D. (2011). *Elements of grading: A guide to effective practice.* Bloomington, IN: Solution Tree Press.

Reeves, D. (2016). *Elements of grading: A guide to effective practice* (2nd ed.). Bloomington, IN: Solution Tree Press.

Resnick, L. B., & Zurawsky, C. (Eds.). (2006). Do the math: Cognitive demand makes a difference. *Research Points: Essential Information for Education Policy, 4*(2), 1–4. Accessed at www.aera.net/Portals/38/docs/Publications/Do%20the%20Math.pdf on July 10, 2017.

Richardson, K. (2012). *How children learn number concepts: A guide to the critical learning phases.* Bellingham, WA: Math Perspectives Teacher Development Center.

SanGiovanni, J. J., Katt, S., Knighten, L. D., & Rivera, G. (2022). *Answers to your biggest questions about teaching elementary math.* Thousand Oaks, CA: Corwin Press.

Schoenfeld, A. H. (1985). *Mathematical problem solving.* Orlando, FL: Academic Press.

Schuhl, S. (2021). Working together to ensure all students learn mathematics. In S. V. Kramer (Ed.), *Charting the course for collaborative teams: Lessons from priority schools in a PLC at Work®* (pp. 131–147). Bloomington, IN: Solution Tree Press.

Schuhl, S. (2023, December 11–13). *Exploring instructional strategies that deepen student learning of mathematics (preK–5)* [Conference Presentation]. Mathematics in a PLC at Work Summit, Las Vegas, NV.

Schuhl, S., Kanold, T. D., Barnes, B., Jain, D. M., Larson, M. R., & Mozingo, B. (2021). *Mathematics unit planning in a PLC at Work, high school.* Bloomington, IN: Solution Tree Press.

Schuhl, S., Kanold, T. D., Deinhart, J., Lang-Raad, N. D., Larson, M. R., & Smith, N. N. (2021). *Mathematics unit planning in a PLC at Work, preK–2.* Bloomington, IN: Solution Tree Press.

Schuhl, S., Kanold, T. D., Deinhart, J., Larson, M. R., & Toncheff, M. (2020). *Mathematics unit planning in a PLC at Work, grades 3–5.* Bloomington, IN: Solution Tree Press.

Schuhl, S., Kanold, T. D., Kanold-McIntyre, J., Chuang, S., Larson, M. R., & Smith, M. (2021). *Mathematics unit planning in a PLC at Work, grades 6–8.* Bloomington, IN: Solution Tree Press.

Schuhl, S., Kanold, T. D., Toncheff, M., Barnes, B., Kanold-McIntyre, J., Larson, M. R., et al. (2024). *Mathematics assessment and intervention in a PLC at Work* (2nd ed.). Bloomington, IN: Solution Tree Press.

She, X., & Harrington, T. (2022). Teaching word-problem solving through tape diagrams. *Mathematics Teacher: Learning and Teaching PK–12, 115*(3), 170–182.

Shute, V. J. (2008). Focus on formative feedback. *Review of Educational Research, 78*(1), 153–189.

Smit, R., Bachmann, P., Blum, V., Birri, T., & Hess, K. (2017). Effects of a rubric for mathematical reasoning on teaching and learning in primary school. *Instructional Science, 45*(5), 603–622.

Smith, M. S., Steele, M. D., & Raith, M. L. (2017). *Taking action: Implementing effective mathematics teaching practices, grades 6–8.* Reston, VA: National Council of Teachers of Mathematics.

Smith, M. S., & Stein, M. K. (1998). Selecting and creating mathematical tasks: From research to practice. *Mathematics Teaching in the Middle School, 3*(5), 344–350.

Smith, M. S., & Stein, M. K. (2011). *Five practices for orchestrating productive mathematics discussions.* Reston, VA: National Council of Teachers of Mathematics.

Smith, M. S., & Stein, M. K. (2018). *Five practices for orchestrating productive mathematics discussions* (2nd ed.). Reston, VA: National Council of Teachers of Mathematics.

Sparks, S. D. (2022, October 24). *Explaining that steep drop in math scores on NAEP: Five takeaways.* Accessed at www.edweek.org/teaching-learning/explaining-that-steep-drop-in-math-scores-on-naep-5-takeaways/2022/10 on January 16, 2024.

STEMscopes Staff. (2022, December 6). All about the concrete representational abstract (CRA) method [Blog post]. *Accelerate the Learning.* Accessed at https://blog.acceleratelearning.com/concrete-representational-and-abstract on July 8, 2024.

Strickland, C. A. (2007). *Tools for high-quality differentiated instruction.* Arlington, VA: ASCD.

Student Achievement Partners. (2015, May 14). *Rigor: Balance conceptual understanding, skills and procedures, and real-world application.* Accessed at https://achievethecore.org/page/1090/rigor on March 19, 2024.

Tomlinson, C. A. (2016). *The differentiated classroom: Responding to the needs of all learners* (2nd ed.). Alexandria, VA: ASCD.

Toncheff, M., Kanold, T. D., Schuhl, S., Barnes, B., Deinhart, J., Kanold-McIntyre, J., et al. (2024). *Mathematics instruction and tasks in a PLC at Work* (2nd ed.). Bloomington, IN: Solution Tree Press.

University of Cambridge. (2016, August 9). *Positive teacher-student relationships boost good behavior in teenagers for up to 4 years.* Accessed at www.sciencedaily.com/releases/2016/08/160809121813.htm on March 19, 2024.

U.S. Department of Education & National Mathematics Advisory Panel. (2008). *Foundations for success: The final report of the National Mathematics Advisory Panel.* Washington, DC: U.S. Department of Education.

Van de Walle, J., Karp, K. S., & Bay-Williams, J. M. (2019). *Elementary and middle school mathematics: Teaching developmentally* (10th ed.). Boston: Pearson.

Venturis, C. (n.d.). Building community in the classroom [Blog post]. *Collaborative Classroom.* Accessed at www.collaborativeclassroom.org/blog/classroom-community on March 19, 2024.

Visible Learning MetaX. (2023, June). *Global research database.* Accessed at www.visiblelearningmetax.com/Influences on March 19, 2024.

Wiggins, G., & McTighe, J. (2011). *The understanding by design guide to creating high-quality units.* Arlington, VA: ASCD.

Wilson, P. (n.d.). *Which one doesn't belong?* Accessed at https://wodb.ca on June 19, 2024.

Wingert, K., & Rhinehart, A. (2016, August). *Talk activities flowchart.* Accessed at https://stemteachingtools.org/assets/landscapes/FullSet_StudentTalkProtocolsandFlowchart1_c.pdf on January 16, 2024.

Would You Rather Math. (n.d.). *Home.* Accessed at www.wouldyourathermath.com on June 19, 2024.

Wright, R. J., Stanger, G., Stafford, A. K., & Martland, J. (2015). *Teaching number in the classroom with 4–8 year olds* (2nd ed.). Thousand Oaks, CA: SAGE.

Wyborney, S. (n.d.). Splat! [Blog post]. *Steve Wyborney.* Accessed at https://stevewyborney.com/2017/02/splat on January 15, 2024.

Zager, T. (2017). *Becoming the math teacher you wish you'd had: Ideas and strategies from vibrant classrooms.* Portland, ME: Stenhouse.

Zike, D. (2003). *Dinah Zike's teaching mathematics with foldables.* Columbia, OH: Glencoe/McGraw-Hill.

Zins, J. E., & Elias, M. J. (2007). Social and emotional learning: Promoting the development of all students. *Journal of Educational and Psychological Consultation, 17*(2–3), 233–255. https://doi.org/10.1080/10474410701413152.

Zwiers, J., Dieckmann, J., Rutherford-Quach, S., Daro, V., Skarin, R., Weiss, S., et al. (2017, February 28). *Principles for the design of mathematics curricula: Promoting language and content development.* Stanford, CA: Stanford University Graduate School of Education. Accessed at https://ul.stanford.edu/sites/default/files/resource/2021-11/Principles%20for%20the%20Design%20of%20Mathematics%20Curricula_1.pdf on March 19, 2024.

# Index

## A
ability grouping, 9
absenteeism, 2
abstract thinking, 97–99
academic achievement
    COVID-19 pandemic and, 1
    high-level-cognitive-demand tasks and, 46–49
    influence of teacher estimates on, 11
academic goal setting, 154
academic standards, xvi
acceleration, 2–4
*Acceleration for All* (Kramer & Schuhl), 45
access, key areas for equitable, 43–44
accountability, 135
accuracy, in feedback, 141
Achieve the Core, 61
*Adding It Up: Helping Children Learn Mathematics* (Kilpatrick, Swafford, & Findell), 15, 98, 129
advocacy, 43–44
agency, 27, 28
Ainsworth, L., 45
algebra tiles, 90
algorithms
    flexibility vs., 69–70
    high-level-cognitive-demand tasks and, 86
    procedural fluency with, 95–98, 109
    unproductive teacher focus on, 11, 16
altruism, 28
analytic rubrics, 160
anchor charts
    of classroom norms, 30–31
    problem solving and, 92, 93
    procedural fluency and, 105–106
    sentence frames and, 134
    Three Es and, 146
Angelou, M., 27
answers, reasonable, 78, 80
Answers First strategy, 105
anxiety, mathematics, 98
application, 16
assessments, 45
    consensus boards and, 124
    creating common, 2, 3–4, 11
    determining learning gaps with, 82–83
    as feedback, 139
    goal setting and, 158, 159
    learning targets and, 46
    preassessments, 59–60
    self-assessment rubric for, 159–160
    student investment and, 5, 154
Assessment Training Institute, 154
Association of State Supervisors of Mathematics (ASSM), 4
attitudes, 27

## B
backward counting strategies, 74–76
Barnes, B., 3, 60–61, 64–66, 86, 115, 130, 141, 158, 165, 174–177
Batelle for Kids, 83–84
Bay-Williams, J., 83
behavior standards, xvi
    norms for student interactions and, 30–32, 39
    sense of belonging and, 28
belonging, 28–29
best first instruction, 12. *See also* Tier 1
Black, P., 139
blame, 22
Boaler, J., 69–70, 106
Bright, G. W., 141
Bruner, J. S., 97
Buffum, A., 11, 12, 17, 43, 44
Building Thinking Classrooms Facebook group, 123
*Building Thinking Classrooms in Mathematics, Grades K–12: 14 Teaching Practices for Enhancing Learning* (Liljedahl), 123
Burns, M., 70

## C
Cahill, J., 123
calculators, 143
Cardetti, F., 123
cardinality, 72, 76
Cardona, M., 1
card sorts, 107, 109, 110–111
Casa, T. M., 123–124
*Catalyzing Change in Early Childhood and Elementary Mathematics: Initiating Critical Conversations* (NCTM), 11, 99–100
centers, 66–67
chant reads, 119
Chappuis, J., 154
Chenoweth, K., 2
Choppin, J. M., 123
classroom environment
    physical arrangement and, 31, 33, 34–35
    student discourse and, 136
cognitive-demand-level task analysis, 179–180
Cohen, J., 123
coherence maps, 46, 61, 62–64
collective teacher efficacy, 11
Common Core Standards Writing Team, 56
communities of learners, 26–41
    characteristics for mathematics students and, 29–30
    environment for, 31, 33, 34–35
    *how* of building, 30–38
    learning from mistakes and, 33–38
    norms for student interactions in, 30–32, 39
    *why* and *what* of building, 27–30
conceptual knowledge, 4
    definition of, 16
    procedural fluency and, 97
conceptual understanding, 15–16
concrete models, 90, 98
    procedural fluency and, 102, 103
concrete thinking, 97–99
consensus boards, 124, 125
conservation of number, 72
*Continuing the Journey: Mathematics Learning 2021 and Beyond* (NCTM, NCSM, & ASSM), 43–44, 60
Counting Off the Decade strategy, 75
COVID-19 pandemic
    absenteeism and, 2
    performance differences since, 1
    urgency of mathematics learning and, 43–44
CRA method, 97–109
critical thinking, 116
Cruz, L. F., 11, 12, 17, 44
culture. *See also* communities of learners
    collaborative, 2
    feedback and, 141

of learning, 4, 7–8
student discourse and, 130
*Curriculum and Evaluation Standards for School Mathematics* (NCTM), 43

**D**

Danielson, C., 120
data analysis protocols, 173–177
decimals, 74, 76
Deinhart, J., 61, 64–66, 141
desk arrangements, 31, 33, 34–35, 136
Dewey, J., 154
Dieckmann, J., 118
Dienhart, J., 61, 64–66, 86, 115, 130, 158
differentiated practice, 106
differentiation, 46–52
learning progressions and, 59
Dimich, N., 11, 12, 17, 44, 140, 142, 153
distributive property, 104, 105
dot images, 72–73
dry-erase desks, 143
DuFour, R., 2, 171

**E**

educational labels, xvii
*Elementary and Middle School Mathematics: Teaching Developmentally* (Van de Walle, Karp, & Fay-Williams), 95, 97
empowerment, 152–169
endurance, 45
*Engage in the Mathematical Practices: Strategies to Build Numeracy and Literacy With K–5 Learners* (Norris & Schuhl), 105, 145–146
English learners (ELs), xvii, 116
equations, solving, 102
equity, 43–44, 172–173
access to grade-/course-level content and, 43–44, 46
identifying essential standards and, 15
mathematical language and, 116
number talks and, 78
performance differences and ethnicity and, 1
Erkens, C., 140, 142, 153
error analysis, 36–38, 149
estimation, 78, 80
Over or Under strategy for, 106
*Every Student Can Learn Mathematics* series, 3–4, 171
four critical questions and, 11
standards classification in, 45
Every Student Succeeds Act, 60
expectations, xv, 157
belonging and, 28
communities of learners and, 30
creating clear and concise, 30–32, 39
feedback on, 141
grade-level, 43, 54, 56, 59
identifying and teaching, 27
key fact and procedural fluency, 95–96
mathematical language and, 116, 119–120
proficiency scales and, 163–164
rocket, 31, 32
rubrics and, 158
for student interactions, 27, 30–33, 39
student tracking and, 165
team discussion of, 10
extensions, 12–13. See also Tier 2
of knowledge, 56

**F**

fact fluency, 95, 96, 98, 99–100
fairness, in feedback, 141
Fay-Williams, J. M., 95, 97
feedback, 5, 138–150
anchor charts in, 105
considering next steps with, 150
definition of, 139
effectiveness of, 139–140
error analysis and, 149
FAST, 141
highlighters and, 148
how of, 142–150
learning teams and, 148
proficiency scales and, 164
Provide the Answers and, 146–148
rubrics and, 161
strategies for, 142–150
student investment and, 154, 155
student questions and, 149
task sorting and, 148
Three Es for, 145–146
types of, 140
why and what of, 139–141
Findell, B., 16
fixed mindset, 141
flex days, 14
flexibility, 69
flexible grouping, 135–136, 148
Flippity, 135
folded scratch paper, 144–145
forward counting strategies, 74–76
foundations in mathematics, 4–5
connecting to prior knowledge, 56–67
number sense, 68–80
problem solving, 82–93
procedural fluency, 94–111
teaching grade-/course-level content, 42–55
fractions, 74
counting with pattern blocks, 75–76
*Framework for 21st Century Learning* (Batelle for Kids), 83
Fredrick, W. C., 97–98
Fuchs, L. S., 97–98

**G**

games, 66–67, 98
gatekeepers, xv
mathematics as, 9
Gates, B., 150
Give Me a Quantity strategy, 72
goal cards, 156–158
goal setting, academic, 154, 158, 159
Gojak, L. M., 33
Gonzalez, J., 161
grade-level learning, 42–55
considering next steps for, 54–55
differentiation across, 46–52
essential standards and, 45–46
focus on, 6
high-level-cognitive-demand tasks and, 46–49
how of teaching, 45–54
leveled tasks and, 49, 52–54
number sense, 70–71
procedural fluency and, 95–99
why and what of teaching, 43–44
grading practices, 11
graphic organizers, 87–89, 123–124, 126
graphing functions, 102–103
graph paper, 143
Gray, E. M., 69
groups
flexible grouping, student discourse, and, 135–136
norms for working in, 30–32, 39
growth mindset, 141
growth model, 60
guiding coalitions, 11–13

**H**

Hanushek, E. A., 1–2
Hattie, J., 4, 11, 84, 136, 139–140
*The Hechinger Report* (Marcus), 1
Heimberg, J., 120
high-level-cognitive-demand tasks, 179–180
differentiation plans for, 50–52
problem solving and, 86
student discourse and, 131
highlighting, 148
holistic rubrics, 159–160
How Full? How Empty? strategy, 74, 75
Humphreys, C., 78

**I**

identity, 27, 28
individual accountability, 135
"Inside the Black Box" (Black & Wiliam), 139
instruction, 171–172
high-quality, 17, 22
meaningful, 4
instruction planning
learning progressions and, 59
unit plans and, 11
interventions, 12–13. See also Tier 2
determining necessary, 46
scheduling time for, 12, 13–14
template for planning, 22, 23–24
investment. See student investment

**J**

jigsaw strategy, 136–137
jobs, mathematics skills required for, 1–2
journals, mathematics, 123–124
Joyner, J. M., 141
just-in-time supports, 60–61
Just Right Numbers strategy, 105–106

**K**

Kanold, T. D., 3, 14, 60–61, 64–66, 86, 115, 130, 141, 158, 165, 174–177
Kanold-McIntyre, J., 3, 14, 60–61, 64–66, 86, 115, 130, 141, 158, 165, 174–177
Karp, K., 83, 95, 97
Kenney, H. J., 97
Kilpatrick, J., 98, 129
Klausmeier, H. J., 97–98
knowledge
conceptual, 4
prior (See prior knowledge)
Known Ten strategy, 100–101
Kramer, S. V., 45
KWL charts, 124–125

**L**

labels. See educational labels
language, xvii
English learners and, 116
mathematical, 5, 114–126
Larson, M. R., 3, 14, 60–61, 141, 165, 174–177
learning
acceleration of, 2–4

capturing and archiving, 171–172
collective responses to, 2, 3
culture of, 4, 7–8
differentiating, 46–49
feedback and, 138–150
focus on, 2
from mistakes, 33–38
passive, 83
shared, 101
strategies for teams focused on, 9–13
teacher influences on, 11
teams focused on, 9–24
learning goals, 150
learning outcomes, xvi, 2
belonging and, 28–29
learning progressions, 2, 14–15, 56–59
addition, 56, 58
connecting to prior knowledge and, 61, 62
definition of, 56
goal cards and, 156–158
procedural fluency and, 100
proficiency scales and, 163–164
learning targets, 46, 47–48
learning teams, 148
learning variances, 2, 3, 8–9
making up for, 42–43, 55–56
lesson design, 26–27
discourse routines and, 130
prior knowledge routines and, 61, 64–66
leveled tasks, 49, 52–54
leverage, 45
Levitt, S. D., 139
Lewis, J., 9
Liljedahl, P., 123, 135, 143

# M
magnitude, 70
Malone, J., 11, 12, 17, 44
manipulatives, 143
Many, T. W., 45
Marcus, J., 1
Martland, J., 74
Marzano, R. J., 4, 139, 154, 159–160
mathematical language, 5, 114–126
consensus boards and, 124, 125
considering next steps with, 125
definition of, 115
graphic organizers and, 124–125, 126
*how* of using, 118–125
mnemonics, 114–115
Share-Trade routine for, 122
strategies for, 118–125
in supporting mathematics understanding, 118

Which One Doesn't Belong? strategy, 120, 121
*why* and *what* of using, 115–118
word walls and, 119–120
Would You Rather? strategy for, 120–121
mathematical literacy, 116
mathematics
coherence map for, 46
foundations in, 4–5
meaningful, 84
performance declines in, 1–2
rigorous, 15–17
mathematics anxiety, 98
*Mathematics Assessment and Intervention in a PLC at Work, Second Edition* (Schuhl, Kanold, Toncheff, Barnes, Kanold-McIntyre, & Larson), 3, 14, 46, 158, 174–177
Mathematics Common Core State Standards, 56
*Mathematics Instruction and Tasks in a PLC at Work, Second Edition* (Toncheff, Kanold, Schuhl, Barnes, Dienhart, & Kanold-McIntyre), 3, 45, 46, 61, 66, 86, 115, 130
*Mathematics Unit Planning in a PLC at Work* series, 15, 22, 45, 117–118
Math-O, 98
Mattos, M., xv–xvii, 11, 12, 17, 44
memorization, 99–100
mistakes, as learning tools, 33–38
error analysis and, 149
My Favorite Mistake strategy and, 106–107, 108
models. *See* concrete models; pictorial models
motivation, 28, 33
MTSS. *See* multitiered system of supports (MTSS)
multitiered system of supports (MTSS), xv. *See also* response to intervention (RTI)
three tiers of, xv–xvi
My Favorite Mistake strategy, 106–107, 108

# N
name and number sort, 71–72
Narayana Murthy, N. R., 115
National Assessment of Educational Progress (NAEP), 1
National Center for Education Statistics, 116

National Council of Supervisors of Mathematics (NCSM), 4
National Council of Teachers of Mathematics (NCTM), 4, 11, 43, 44, 69, 95, 116–117, 129, 132
National Research Council, 98
negative integers, 76
No Child Left Behind, 60
noncounting strategies, 100
norms, for student interactions, 30–32, 39
Norris, K., 85, 105, 116, 133
*note taking,* 123
*not yet,* the power of, 33, and 36
number lines, 56, 76–78
number sense, 4–5, 68–80
considering next steps for, 80
definition of, 69
estimation and reasonable answers and, 78, 80
forward and backward counting strategies for, 74–76
by grade band, 70
*how* of developing, 71–80
number lines and, 76–78
number talks and, 78, 79
strategies for developing, 71–80
subitizing routines for, 72–75
*why* and *what* of developing, 69–71
number talks, 78, 79, 98
numeracy, early, 71–72

# O
one-to-one correspondence, 72, 76
Osterman, K. F., 27
Over or Under Estimation strategy, 106

# P
parent functions
transforming, 103
word walls on, 119
Parker, R., 78
Parrish, S., 78
pattern blocks, 75–76
personal-level feedback, 140
physical environment, 31, 33
pictorial models, 98
procedural fluency and, 104, 105
preassessments, 59–60
preventions, 11–12. *See also* Tier 1
*Principles and Standards for School Mathematics* (NCTM), 69, 84
*Principles to Actions: Ensuring Mathematical Success for All* (NCTM), 11, 84, 116–117

prior knowledge, 4, 56–67
centers and games for, 66–67
connecting to daily lessons, 45–46
considering next steps with, 67
*how* to connect to, 59–67
just-in-time supports and, 60–61
learning progressions and, 61, 62
number talks and, 78
preassessments and, 59–60
routines for, 61, 64–66
Tier 2 interventions and, 12–13
*why* and *what* of, 56–59
problem solving, 5, 82–93
considering next steps for, 92
definition of, 83, 84
graphic organizers in, 125
*how* of focusing on, 85–92
strategies for developing, 85–92
student discourse and, 129
tasks in, 84–85
visuals and tools for, 87–90
*why* and *what* of, 83–85
word problems and, 91–93
problems without numbers, 91–93
procedural fluency, 5, 16–17, 94–111
card sorts and, 107, 109, 110–111
considering next steps for, 109
CRA method and, 97–99
definition of, 95
existing pictorial models for, 104, 105
flexible strategies vs. memorization in, 99–100
*how* of developing, 99–109
Just Right Numbers strategy and, 105–106
key skills in, 95, 96
My Favorite Mistake for, 106–107, 108
Over or Under Estimation and, 106
strategies for, 99–109
*why* and *what* of developing, 95–99
process feedback, 140
Professional Learning Communities (PLCs) at Work, xvii
four critical questions in, 3, 11, 14, 22, 171
mathematics in, 2–4
three big ideas of, 2
proficiency scales, 163–164
protocols

data analysis, 173–177
for differentiation across grade levels, 46–52
rigorous mathematics and, 17–21
three reads, 91–93
Provide the Answers strategy, 146–148
purposeful questioning, 132–133

## Q

questioning techniques, 36, 129
    feedback and, 149
    planning and, 171, 172
    purposeful, 132–133

## R

race and ethnicity, performance differences and, 1
Raith, M. L., 130
readiness, 45
*REAL* acronym for standards, 45
reasonable answers, 78, 80
reasoning, 85, 129
Reeves, D. B., 45
reflection, 154
    on belonging, 28–29
    on communities of learners, 39
    feedback and, 150
    on grade-level teaching, 54–55
    holistic rubrics and, 159–160
    on number sense, 80
    on prior knowledge, 67
    on problem solving, 92
    on procedural fluency, 109
    proficiency scales and, 164
    questioning to encourage, 132
reinforcement, xvi–xvii
    intensive, 13
remediation, 9
representational thinking, 97–99
response to intervention (RTI), xv–xvii
    role of collaborative teams in, 4, 11
    three tiers of, xv–xvi
responsibility, collective, 2
results orientation, 2
retention, 9
revising vs. redoing, 145
Rhinehart, A., 122
Richardson, K., 69, 78
rigor, mathematical, 15–17
    definition of, 15
    procedural fluency and, 98
    protocols for, 17–21
rocket expectations, 31, 32
routines, 11
    mathematical language and, 118–125
    prior knowledge, 61, 64–66
    student discourse and, 130, 131

subitizing, 72–75
RTI. *See* Response to Intervention (RTI)
rubrics, 158–161
    for high-quality Tier 1 and Tier 2 strategies, 9, 10
    Teacher Team Rubric: Tier 2 Mathematics Intervention Program, 14

## S

safe environments, 33, 150
*Salute!* card game, 66–67
scaffolding, 49
    just-in-time supports and, 60
    learning progressions and, 59
schedules and scheduling
    for Tier 1, 14
    for Tier 2, 12, 13–14, 15
Schimmer, T., 140, 142, 153
Schoenfeld, A. H., 83
Schuhl, S., 3, 11, 12, 17, 44, 45, 60–61, 64–66, 85, 86, 92, 105, 115, 116, 130, 133, 141, 158, 165, 174–177
scratch paper, folded, 144–145
See, Say, Write activity, 72–74
self-efficacy, 27, 28, 158–160
self-regulation, 154
    rubrics and, 158–161
self-regulation feedback, 140
sensemaking, 85
sentence frames, 133–135
shared accountability, 135
Share-Trade protocol, 122
Shute, V. J., 140
single-point rubrics, 160–161, 162
Skarin, R., 118
skills
    job requirements for, 1
    reteaching, 60
Smith, M. S., 86, 130, 132
specificity, in feedback, 141
Splat! images, 73–74
Stafford, A. K., 74
standards
    access to grade-level, xvi
    coherence among, 56
    essential, data analysis protocol on, 176–177
    essential, determining, 9, 43
    essential, instruction on, 14–15
    essential, planning interventions for, 22, 23–24
    essential, teaching, 44, 45–49
    important-to-know, 45
    interventions focused on, 12–13
    making sense of, 2

    on mathematical language, 116–117
    nice-to-know, 45
    on problem solving, 84
    *REAL* acronym for, 45
    on student discourse, 129–130
    success criteria for, 161–163
    teaching grade-/course-level content and, 43–44
Standards for Mathematical Practice, 116–117
Stanger, G., 74
Steele, M. D., 130
Stein, M. K., 86, 132
Stiggins, R., 154
strategies
    connecting and comparing, 104–105
    for connecting to prior knowledge, 59–67
    for developing number sense, 71–80
    how to use, 4–5
    ineffective, 9
    for mathematical language, 118–125
    problem solving and, 85–92
    for procedural fluency, 99–109
    rubric for high-quality Tier 1 and Tier 2, 9, 10
    for student discourse, 132–137
    for student investment, 156–168
    for teams focused on student learning, 9–13
    vertical progression of, 101–103
Stronger and Clearer Each Time strategy, 122
student discourse, 5, 128–137
    considering next steps with, 137
    differentiation and, 46–49
    engagement and, 128–129
    flexible grouping for, 135–136
    How Full? How Empty? and, 74, 75
    *how* of, 132–137
    jigsaw strategy for, 136–137
    mathematical language and, 116–117
    purposeful questioning and, 132–133
    rights and responsibilities for, 30–31
    roles for, 135
    sentence frames and, 133–135
    strategies for, 132–137
    *why* and *what* of, 129–131
student engagement, 5, 13, 26–27
    feedback and, 138–150
    investment and, 152–169

    mathematical language and, 114–126
    student discourse and, 128–137
    in thinking, 123
student investment, 5, 152–169
    considering next steps with, 169
    definition of, 153
    goal cards and, 156–158
    goal setting and, 158, 159
    *how* of, 156–168
    proficiency scales and, 163–164
    rubrics and, 158–161
    strategies for, 156–168
    student surveys and, 165
    student trackers and, 165, 166–169
    success criteria and, 161–163
    *why* and *what* of, 153–155
student surveys, 165
student thinking, 2
    CRA method and, 97–99
    critical, 116
    determining trends in, 173–174
    engagement in, 123
    problem-solving reasoning and, 5
    showing the mathematics and, 142–145
student trackers, 165, 166–169
student work, analyzing, 2, 171–172
student work protocols, 175
subitizing routines, 72–75
success, xv
success criteria, 161–163
Swafford, J., 98, 129

## T

*Taking Action: A Handbook for RTI at Work* (Mattos, Buffum, Malone, Cruz, Dimich, & Schuhl), 11
talking sticks, 134–135
Tall, D. O., 69
task analysis guide, 179–180
task-level feedback, 140
task sorting, 148
teachers
    beliefs of, productive/unproductive, 11
    collective efficacy of, 11
    continuous learning for, 2–4
    as gatekeepers, xv
    mathematical language use by, 117
    shortages of mathematics, 2
    tracking, 9
Teacher Team Rubric: Tier 2 Mathematics Intervention Program, 14

teaching practices, 84
teams, collaborative, 2, 9–24, 171–172
    actions to avoid and considered by, 181–182
    building a community of learners and, 26–61
    considering next steps for, 22
    high-quality instruction and, 17, 22
    how to work as, 13–22
    instruction on essential learning standards and, 14–15
    mathematical language support by, 116, 117–118
    rigorous mathematics and, 15–17
    in RTI, 4
    time in the day for Tier 2 and, 13–14
    *why* and *what* of, 9–13
templates
    goal setting, 158
    name and number sort, 71
    for planning Tier 2 interventions, 22, 23–24
    teacher team planning and leveled tasks, 52–53
ten frames, 54, 72, 97, 100–101, 142–143
thinking. *See* student thinking
Three Es strategy, 145–146
three reads protocol, 91–92, 93
Three Stay, One Stray activity, 135–136
Tier 1, xv–xvii
    connecting to prior knowledge in, 59
    CRA method in, 97, 99
    guiding coalitions in, 12
    high-quality instruction for, 17, 22
    importance of, xvii
    number sense in, 70–71
    preventions in, 11–12
    problem solving in, 85, 86
    rubric for high-quality strategies in, 9, 10
    scheduling for, 15
    strategy use in, 4–5
    teaching grade-level standards in, 44
Tier 2, xv–xvii
    connecting to prior knowledge in, 59
    CRA method in, 97–99
    high-quality instruction for, 17, 22
    importance of, xvii
    number sense in, 70–71
    problem solving in, 85, 86
    rubric for high-quality strategies in, 9, 10
    scheduling time for, 12, 13–14
    strategy use in, 4–5
    success metrics for, 13
    teaching grade-level standards in, 44
Tier 3, xv–xvii, 13
timeliness, of feedback, 141
Timperley, H., 140
Tomlinson, C. A., 46
Toncheff, M., 3, 14, 60–61, 64–66, 86, 115, 130, 141, 158, 174–177
tracking, 9, 44

## U

uninterrupted instruction, 12
unit plans, 11
U.S. Department of Education, 97–98, 115

## V

Van de Walle, J., 83, 84, 95, 135
virtual tools, 78
visuals and tools
    problem solving and, 87–90
    procedural fluency and, 101–103
vocabulary, xv–xvi, 115

## W

Which One Doesn't Belong? strategy, 120, 121
whiteboards, 143
Wiliam, D., 139
Wilson, P., 120
Wingert, K., 122
word problems, 91–93
word walls, 119–120
Would You Rather? strategy, 120–121
Wright, R. J., 74, 98
writing
    mathematics journals, 123–124
    Stronger and Clearer Each Time, 122
Wyborney, S., 73

## Y

YouCubed, 70

## Z

Zager, T., 33
Zwiers, J., 118

**Mathematics Assessment and Intervention in a PLC at Work®, Second Edition**
*Sarah Schuhl, Timothy D. Kanold, Mona Toncheff, Bill Barnes, Jessica Kanold-McIntyre, Matthew R. Larson, and Georgina Rivera*
Build collective teacher efficacy by using the Mathematics in a PLC at Work™ common assessment process. New and enhanced second edition tools support your collaborative teams in developing Tier 2 interventions as part of teacher and student reflection and action.
**BKG146**

**Mathematics Instruction and Tasks in a PLC at Work®, Second Edition**
*Mona Toncheff, Timothy D. Kanold, Sarah Schuhl, Bill Barnes, Jennifer Deinhart, Jessica Kanold-McIntyre, and Matthew R. Larson*
Build collective teacher efficacy and students' mathematical thinking using the Mathematics in a PLC at Work™ lesson-design process. PreK–12 teacher teams will learn new and enhanced research-affirmed lesson-design elements and how to efficiently elicit high levels of student engagement and self-efficacy.
**BKG147**

**Mathematics Homework and Grading**
*Timothy D. Kanold, Bill Barnes, Matthew R. Larson, Jessica Kanold-McIntyre, Sarah Schuhl, and Mona Toncheff*
Rely on this user-friendly resource to help you create common independent practice assignments and equitable grading practices that boost student achievement in mathematics. The book features teacher team tools and activities to inspire student achievement and perseverance.
**BKF825**

**Mathematics Coaching and Collaboration**
*Timothy D. Kanold, Mona Toncheff, Matthew R. Larson, Bill Barnes, Jessica Kanold-McIntyre, and Sarah Schuhl*
Build a mathematics teaching community that promotes learning for K–12 educators and students. This user-friendly resource will help you coach highly effective teams within your PLC and then show you how to utilize collaboration and lesson-design elements for team reflection, data analysis, and action.
**BKF826**

**Taking Action, Second Edition**
*Mike Mattos, Austin Buffum, Janet Malone, Luis F. Cruz, Nicole Dimich, and Sarah Schuhl*
The second edition of the bestseller *Taking Action* delves deeper into the essential actions needed to create a highly effective multitiered system of supports. New recommendations and tools are included to better target assessments, engage students, and proactively address resistance.
**BKG136**

**Solution Tree | Press**
*a division of Solution Tree*

Visit SolutionTree.com or call 800.733.6786 to order.

# Global PD teams
### Collaborative Learning for School Improvement

# Quality team learning **from authors you trust**

Global PD Teams is the first-ever **online professional development resource designed to support your entire faculty on your learning journey.** This convenient tool offers daily access to videos, mini-courses, eBooks, articles, and more packed with insights and research-backed strategies you can use immediately.

**GET STARTED**
SolutionTree.com/**GlobalPDTeams**
800.733.6786

Solution Tree's mission is to advance the work of our authors. By working with the best researchers and educators worldwide, we strive to be the premier provider of innovative publishing, in-demand events, and inspired professional development designed to transform education to ensure that all students learn.

The National Council of Teachers of Mathematics advocates for high-quality mathematics teaching and learning for each and every student.